FROM THE SEASHORE TO THE SEAFLOOR

PAINTINGS BY **Peggy Macnamara**

TEXT BY **Janet Voight**

WITH A FOREWORD BY **David Quammen**

Published in association with the Field Museum

FROM THE SEASHORE TO THE SEAFLOOR

AN ILLUSTRATED TOUR OF SANDY BEACHES, KELP FORESTS, CORAL REEFS, AND LIFE IN THE OCEAN'S DEPTHS

THE UNIVERSITY OF CHICAGO PRESS

CHICAGO AND LONDON

The University of Chicago Press, Chicago 60637
The University of Chicago Press, Ltd., London
© 2022 by The University of Chicago
Published 2022
Printed in Canada

31 30 29 28 27 26 25 24 23 22   1 2 3 4 5

ISBN-13: 978-0-226-81766-8 (cloth)
ISBN-13: 978-0-226-81770-5 (e-book)
DOI: https://doi.org/10.7208/chicago/9780226817705.001.0001

♾ This paper meets the requirements of ANSI/NISO Z39.48-1992
(Permanence of Paper).

Library of Congress Cataloging-in-Publication Data

Names: Voight, Janet, author. | Macnamara, Peggy, artist. | Quammen,
   David, 1948– writer of foreword.
Title: From the seashore to the seafloor : an illustrated tour of sandy
   beaches, kelp forests, coral reefs, and life in the ocean's depths /
   paintings by Peggy Macnamara ; text by Janet Voight ; with a fore-
   word by David Quammen.
Description: Chicago : The University of Chicago Press, 2022. |
   "Published in association with the Field Museum."
Identifiers: LCCN 2022001409 | ISBN 9780226817668 (cloth) |
   ISBN 9780226817705 (ebook)
Subjects: LCSH: Marine animals. | Marine ecology. | Marine animals—
   Pictorial works. | Marine ecology—Pictorial works.
Classification: LCC QH541.5.S3 V65 2022 | DDC 577.7—dc23/
   eng/20220126
LC record available at https://lccn.loc.gov/2022001409

For Bridgid O'Connor Creevy

and Dorothea Bridgid Peterson

for keeping me connected

to a better world

CONTENTS

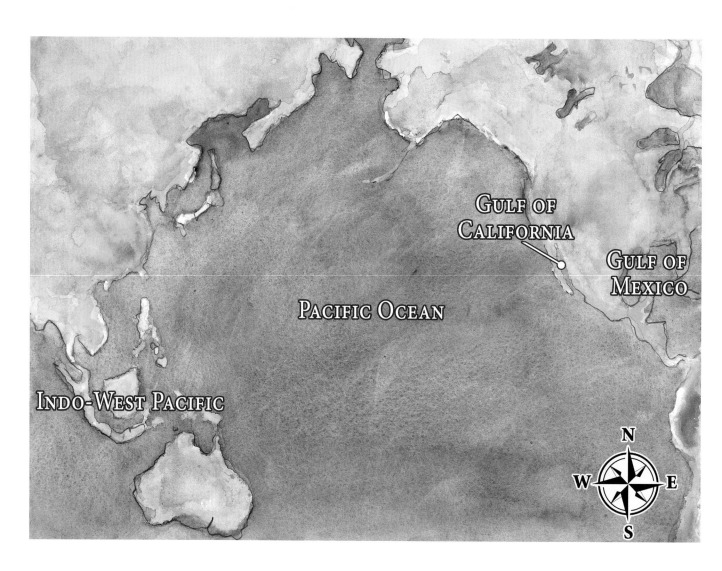

FOREWORD

Of all the improbable marine creatures that populate Earth's oceans, my own favorites have long been the octopuses. They constitute an order of cephalopods, Octopoda, with about three hundred named species, and they are graced by a richly oxymoronic combination of traits—some quite alien to our human perspective, others that seem spookily familiar.

Octopuses. Their bodies are soft, except where (in the beak and between the eyes) they are hard. Their arms are eight. They have no legs but they can walk. They can also, at least certain of them, change color at will. They move slowly except when they move fast. They are muscular, strong, and grabby without benefit of a skeleton. They can squeeze through small gaps, but their brains are big. And they gaze out at the world through a pair of complex eyes structured much like the vertebrate eye—each with an iris, a circular lens, a vitreous cavity, a light-sensitive retina—but produced (probably) by convergent evolution. All this and more explains why I'm so glad to see octopuses lovingly featured among the images and words of the book you now hold.

To speak personally for a moment: When I first got interested in octopuses and wrote a little something about them more than thirty-five years ago, I was blessed with an advantage. I had a friend named Janet Voight, then a graduate student at the University of Arizona, in Tucson, doing her doctorate on Pacific pygmy octopuses of the Gulf of California. Janet was keeping a few of these delicate little things temporarily captive, living in tanks, in the basement of a departmental building, and she indulged me in a visit to see them and ogle. They were mostly too shy to ogle back (or else they simply had better things to do). But if for them it was a nonevent, for me it was memorable. Those close observations gave me a sense of octopushood I could hardly otherwise have gotten. I wrote my little piece, and our paths diverged, Janet and octopuses going one way, me another.

Ten years later, Dr. Voight got her fifteen minutes of glaring, unwanted Warholian fame—talk radio interviews, television, the Associated Press, *Playboy* magazine—because she had coauthored a paper in *Nature* describing an extraordinary encounter: mating between two male deepwater octopuses of different species as captured by video camera from the research submersible *Alvin*, over eight thousand feet down in the eastern Pacific. By that time Janet was based in Chicago, a curator at the Field Museum, and I was still writing a monthly column about natural science. Catching wind of the octopoid ruckus, spotting her name, I was able to prevail on her friendship again. She trusted me to write the more complex ecological story behind the lurid, confused cartoon being peddled in the other media. Never mind, now, the subject of that publicity boomlet. The point for Janet, and for me following her lead, was that the ordinary facts of octopus biology are more than wondrous enough, while the anomalies are just distraction.

This is true also, of course, for the creatures with

whom octopuses share habitat, the fishes great and small, the marine mammals and reptiles, the other mollusks and the jellyfishes and the tube worms and all else. And that's what this book is about. That's what it says, in both its visual and verbal modalities: Look closely, dear people. Look with sympathy and fascination and awe. Look on these majesties of marine life, read about them, learn something about them—and be grateful you were born on the blue planet.

Finally, it's just coincidence, but happy coincidence, that another old friend of mine, the artist Peggy Macnamara, also of the Field Museum, has collaborated with Janet to produce such a deeply observed and exquisitely layered book. Savor it. The minds and the eyes of these two journeying women will take you places you haven't been.

DAVID QUAMMEN

INTRODUCTIONS

SCIENTIST INTRODUCTION

Oceans cover 70 percent of our planet, and our knowledge of them is woefully incomplete. No book of this length can show you all known marine life. We have chosen to take you to fourteen very special habitats from the ocean shore to its floor where, with one exception, I have done research. Our goal is to give you a biologist's view of life and the environment and to explain just a bit about how precious and delicate both are.

I try to make these alien worlds real through words; Peggy makes the animals come alive through watercolors. She uses just a bit of text within and between the chapters to explain her approach and technique in the book. I explain my approach and technique in a couple of chapters about life aboard a research cruise and working in *Alvin*, a submersible that can carry two scientists and the pilot to depths of 15,000 feet (4,500 me-

ters). What we have chosen to describe and depict is personal for both of us. We hope caring for the oceans becomes personal for you too.

I've always been fascinated by animals, eager to learn how they live and how their parts work. Growing up in Iowa, my career path seemed unlikely. My parents hadn't gone to college, and they had no idea that their little girl would grow up to be a scientist or even that such a thing was possible. In that era women were not expected to become scientists; I was told perhaps I could be a nurse or a teacher. When I entered college at Iowa State University, it was as if the world opened up to me; I was so happy to have the opportunity to study anything I wanted (and physics). As a college student in the mid-1970s, I managed to land an internship at the Iowa Coal Project. It was the 1970s energy crisis, and the state of Iowa was looking at whether its coal resources could be exploited safely. As an intern I was charged with monitoring small mammal populations, and in my spare time I looked at birds. I knew

the opportunity was special because it allowed me to generate data and record observations. With some extra effort, this research resulted in published scientific papers with me as an author—an extremely rare thing for female undergraduates.

After college I met a senior, very well-respected scientist, Dr. Annette Fitz, who had established herself on her own merit as a clinician and was a member of the teaching faculty at the University of Iowa Hospitals. For the first time, I identified with a scientist, one who was a woman and unmarried and didn't seem to care about that! She eventually told me I should go to graduate school because I had the ability do my own research instead of simply helping with someone else's.

I had no idea of where to go or how to apply, but I was ready. I figured it out by myself, kind of by trial and error (with lots of errors). I applied to five schools, got into two, and chose the better program that accepted me with financial support. It took me a while to get back into being a student at the University of Arizona after four years of working nine to five. But

I found my footing when I took a marine field course as a lark, studying the northern Gulf of California, which was only about 214 miles (about 340 kilometers) from the university. I had grown up in Iowa, so I didn't know much about the oceans, and I promptly fell in love. Over time, I became an expert on octopods; these mollusks are like us in many ways but are nearly as different as they can be and still be an animal. They seem to be intelligent, but their brains are said to be in their arms. Their eyes work like ours, but rather than changing the shape of the lens to look at something close-up, they change the shape of their eyeball. Most octopods don't chew their food; they inject their prey with venom, wait a little bit, and then after the venom has made soup of the prey's insides, slurp up the mush. Later, as a new curator at the Field Museum in Chicago, I was told I could do fieldwork anywhere in the world. When I imagined a map of the world, all I could see was the deep blue between continents.

How could I get on an oceangoing research ship? I had to figure it out. I contacted people whose publications interested me, and I let them know that I would love to go to sea. I do believe I generated some pity in the deep-sea research community because after a few years of pestering, I was invited to go on a cruise. I have now sailed on twenty-two research cruises, two as chief scientist—or the person who proposes what the group will study, gets the grant, organizes the cruise and who, with the captain, makes all the decisions (except what to have for dinner). I've been on eight dives to as deep as 10,738 feet (3,273 meters) inside the deep ocean research vessel *Alvin*, and I have participated in hundreds of hours of Remotely Operated Vehicle (ROV; basically, a robot commanded by people aboard the ship while the robot is on the seafloor) dives, what we ocean scientists call "lowerings." Among my career highlights are leading a successful three-and-a-half-week cruise with the *Alvin* to hydrothermal vents on the East Pacific Rise, which is a boundary between continents that runs between California and Antarctica. Here, we discovered a long-neglected ecosystem dominated by a species of Stauromedusa, a kind of tall jellyfish that was later named for me: *Lucernaria jane-tae*. There have been other species named for me since then, but the first one is oh so special!

Until very recently, the glories of the depths and the complexities of life under the waves have been hidden from us humans. My career as an explorer of this habitat has been extraordinary. As mammals, we need air, so for most of history we've only been able to skim the ocean's surface on boats or probe the sea's shallow waters for food. The oceans have represented boundaries and infinities; they are both a source of a seemingly endless bounty of food and the easiest place to dispose of garbage. As a scientist who has seen the ocean's depths, I understand they are both expansive and finite. The seas need our protection, especially as they are home to most of the planet's animal life and the most different kinds of animals at the level of phylum.

I've found a way to transcend these waves, and you will too. You can take yourself there in this book, guided by

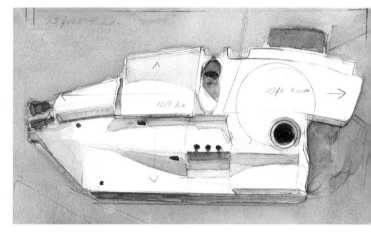

Alvin

me, a deep-sea biologist, and illuminated by extraordinary watercolor images by my friend, the incredibly talented Peggy. We feature many of the ocean's splendors by focusing on familiar yet mysterious marine habitats and their residents. We start where so many people have their first experience with the ocean: the

beach! We move among shallower habitats before heading out to—and into—the open ocean. *Moby-Dick*'s author Herman Melville compared the ocean to a tart; we think you'll agree that the crust is tasty, but you eat a tart for the filling!

We hope the words and images in this book will help you understand our fascination with the oceans and the life they contain. I also hope the story inspires you to follow your own inspiration. You'll need what's called tenacity, or just plain stubbornness. If you find something that calls to you, you, too, can figure out, with lots of effort and persistence, how to make it yours—and to share it with others.

JANET VOIGHT, PHD
Women's Board Associate Curator for Invertebrate Zoology,
Field Museum, Chicago

ARTIST INTRODUCTION

Throughout this book you will see artist notes. Here I will set the stage for these notes and the watercolors that I created for each chapter. My background is not in science; it still baffles me, but the subject matter inspires me. I studied art history in college and graduate school. I once thought artists were born with drawing skills, unique color sense, and lots of ideas. None of this is true. I still wanted to draw and paint even though I possessed none of the skills necessary. Lucky for me, I decided to try anyway. I loved the Renaissance masters and their lengthy training, and so I decided to make their path my own. I drew figures daily for three

to five years. And I had an old-fashioned instructor who taught technique not ideas.

After three years my drawing began to improve, and I went to the Field Museum. Again, I did not have a strong interest in science, but I hoped, with subjects that held perfectly still day after day, that I could correct endlessly and get the drawings right. I didn't know then that mastering the skill of drawing was teaching me color and composition. See, when you are drawing one area of your subject, it must relate to all the other areas of the subject. You are carrying the whole in your head while establishing one area at a time. So when you get to color and you put the first one down, you usually will move around the page to every area where that color appears. Color, like drawing, is about relating parts. With practice, this ability to keep the whole in mind becomes second nature. Then you choose subjects, begin to look at other work to see if you find yourself there, and you are off and running!

When you start a piece, you are usually acting like

a newlywed . . . all excited and sure everything will be great. But as you proceed to the second or third day, trouble inevitably appears. Sleep well and go back at it the next day. Fix the easiest problem to fix, then on to the next, and so on and so on. Sometimes you get lucky and the piece works, but actually who cares . . . you are getting better.

I believe there is much more talent out there than ever gets exposed. The trick to the whole thing is just do it and don't think too much. You will be amazed with what you come up with. Observing and rendering nature is like sitting with a master who subtly guides your every move. You take in nature's nuances and perfection, and you develop your own. Creating will train your instincts, surprise and delight you. It will also help you develop your unique skills. When you paint birds, mammals, plants, and fish, you are "Playing Up"; you are hanging with the best, top of the line, ultimate masters. In that company, you will develop and shine!

The painting techniques that I describe in the art-

ist notes are developments occurring after forty years of painting most days. I hope they help you see that most of the process is made up of recognizing where you went wrong and just fixing it as best you can. Have fun!

PEGGY MACNAMARA

Artist in Residence, Field Museum, Chicago
Associate Professor, School of the Art Institute of Chicago

CHAPTER 1

Sandy Shores

Imagine an endless beach on a summer day with waves gently breaking and rolling up on the shore, stirring the sand as they retreat only to advance again, time after time after time. It's one of the most serene places on earth. For humans, large, air-breathing animals covered with a thick skin, it offers the mix of air and water, rhythmically comforting sounds, and soothing motion. For the ever-so-small animals that live in the sand, however, life is not so idyllic. The wave energy moves individual sand grains, whether they be made of tiny bits of rock that have washed down from the land or the skeletons of tiny mollusks, like tiny snails and bivalves. These grains bounce across the surface and run into anything in their way, including animals living partially buried in the sand. But the waves agitate the sand and stir up morsels of food that sustain life and make living in a sand blaster possible.

Although the beach may seem nearly free of animals other than birds, a diverse array of organisms inhabits the sand. Buried in the sand are clams, their bodies covered by two paired shells that can close tightly to protect them when the need arises. These clams stick out long tubes, their siphons, to take in water and strain it through their gills to filter it. When you look closer, you may see polychaete—segmented and bristly—worms, crustaceans, mollusks, and nemerteans—ribbon worms with beautiful colors but ferocious appetites for other animals. Those animals are all big enough that they burrow into the sand, sometimes forming permanent holes. The sand is

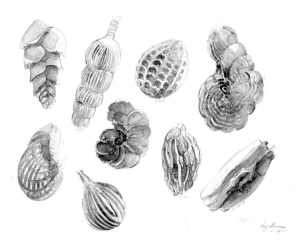

Foraminifera shells. Range from 0.004 to 7.9 inches (100 micrometers to almost 20 centimeters) long.

also home to animals so small they live in the spaces between sand grains.

These, as you might expect, don't really have common names, but they include arthropods (jointed-legged animals like roly-poly bugs that live in the ocean, not just in the soil), nematodes (round worms), polychaetes, and many others. Just about any kind of animal you can think of (and many kinds you never imagined) lives among the sand grains. They are incredibly important because many animals, like sea cucumbers and sipunculans, eat sediment. They don't eat sediment because they like dirt; they eat it to feed on those tiny animals living within or on top of it.

Larger predators abound just under the sand, too. The moon snail, with its coiled shell a sandy-colored globe, emerges to crawl over the sandy surface on an immense extensible foot that not only moves it forward with impressive speed but can propel it into the sand. Murex snails are predators that find and excavate clams that thought they were safely burrowed under the sand. The murex drills a hole through the clam shell and injects the soft parts of the bivalve with a poison that relaxes its muscles so its shells open; the snail then feasts on the body inside. Sand-dwelling sea stars have pointy tube feet that work well in sand, as the round tube feet of typical sea stars would be nearly useless. Even fish live buried in the sand. Garden eels extend only their heads and the front part of their bodies into the current to feed on morsels of food being swept past in the water.

High tide brings more predators who move in to take what they can in the tumult created by the advancing water. Animals torn out of their burrows and stranded during low tide are detritus to be fed on, as are unsuspecting animals too preoccupied with their own dinner to realize they may become someone else's. These animals that advance with the tide are the original marauders; the flood tide turns what had been a placid beach into an area ripe for their depredations. Sharks, stingrays, and bony fishes keep a sharp lookout for sand dwellers not quite covered by sand. Sting rays have flat bodies with a mouth on the bottom. Once a stingray locates a clam, it covers it and extends its mouth to crush the clam. Fishes can pick at exposed siphons or bite material that has drifted with the sand grains.

Turban snail (*Guildfordia* sp.). 5 species. 3–4 inches (8–10 centimeters) across, including the spines.

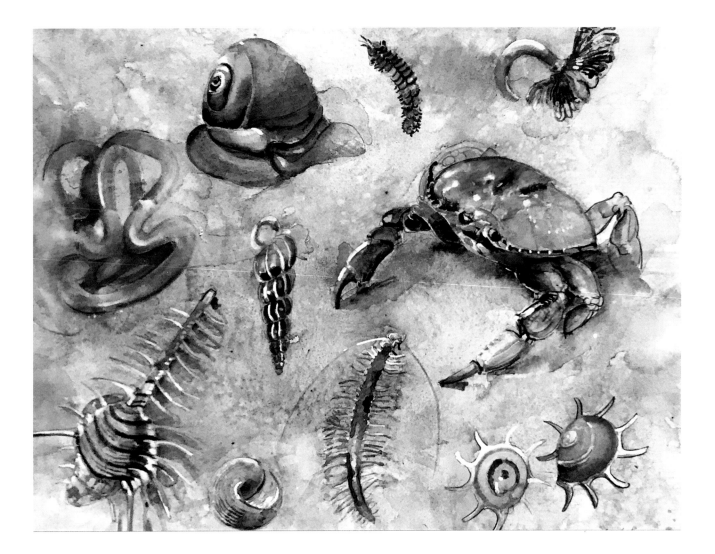

Opposite, clockwise from upper left: Ribbon worm Phylum Nemertea (1,300 species). Range from 0.1 inches (3–4 millimeters) to (rarely) up to 177 feet (54 meters) long. Moon snail (*Neverita lewisii*). 5.5 inches (14 centimeters) long. Polynoid worm Phylum Annelida (868 species). Range up to 4 inches (100 millimeters). Peanut worm Phylum Sipuncula (162 species). Range from 0.1 inches to 28 inches (2.5 to 711 millimeters) long. Generalized crab. Range from 4.5 to 6 inches (11 to 15 centimeters) across. Turban snail (*Guildfordia* sp.). 5 species. 3–4 inches (8–10 centimeters) across, including the spines (*lower right corner*). Polychaete worm Phylum Annelida (over 8,000 species). Range from 0.04 inches (1 millimeter) to 10 feet (3 meters) long. Cone snail Family Conidae (500 species). Range from 0.5 inch (1.3 centimeters) to 8.5 inches (21.6 centimeters) long. Venus comb murex (*Murex pecten*) Indo-West Pacific. 6 inches (15 centimeters). Wentletrap (*Epitonium clathrum*). 1 inch (2.5 centimeters) (*center*).

ARTIST NOTE

This chapter is a strong art lesson. The act of looking in art is more than just a glance. Daily life only requires we take in information. When observing nature and painting, prolonged observation is necessary. This habit leads to discovery in both science and art. Who knew the sand held such innovative, magical life? Patience is an important technique. A week into painting the above images of beach species, I started adding soft washes of the paints manganese blue and titanium white. This way I was able to illuminate the species top second from the left. I was also able to create the illusion of sand with titanium and ochre. If your painting doesn't look right, let it dry (or have its say) and then soften with opaque washes. Lemon yellow and water is a color I used to both cover and soften.

CHAPTER 2

Rocky Shores

As you move away from the sandy beach, you start to encounter scattered rocks. These are probably sand blasted or periodically even completely buried by sand, so they have no detectable animal life. In contrast, when you reach intertidal rocks that are free of encroaching sand, you may find dazzling arrays of animal life, if you know where to look. Because the intertidal zone—the land exposed between the highest high tide and the lowest low—is a hard place to live, animals can't live on the surface of flat rocks; they would be periodically exposed to the burning rays of the sun, to the drying air, and to crashing waves and would only periodically enjoy full immersion in the sea. The animals live in rock crevices and on the underside of rocks.

Most animals in this area are sessile—meaning as adults they can't get up and move around. This means that they don't have to hang on when big waves hit

the area because they are already attached, but it also means that they can't leave when low tide exposes them to the drying effects of air and sunshine.

Being positioned at the land edge of the intertidal zone means you are tough, able to take days without immersion in seawater. Often barnacles are the first animal indicators that you are in the extreme reaches of the intertidal zone. Most other animals tend to live where there is water if not all the time, then at least with fair frequency. Flip over a large rock and you will be amazed at the color riot on the underside. Some of these animals may not on first glimpse look like animals. They are blobs. Tunicates can be bright blobs that look and feel like congealed snot. These animals filter feed, straining food particles from the water. Tunicates contrast with what can be the equally colorful filter-feeding sponges in texture. Sponges are often porous, like the kind kept by the kitchen sink. But look closer, see that hard crust on the rock? It's a moss animal or bryozoan colony. They form hard sheets made up of thousands of individuals. Each is distinct but forms an essential part of the whole.

Anemones can play a big role in the intertidal zone. They extend their tentacles in hopes that something touches them. When it does, the tentacles wrap themselves around their prize and pull it toward the mouth in the center of the animal. Anemones have beautiful colors, and the tentacles of most will feel a bit sticky to you if you touch them. If you had more delicate skin, you would feel a great deal more. It might start as a prickly feeling and become a burning sensation that would go on for hours.

Mobile animals also live on these rocks, but some look as if they are nearly sessile. Animals called chitons or "sea cradles" scrape the rock with the teeth of their conveyer belt–like radula. When one feels threatened, either by you or by an increase in wave action, it uses its large muscular foot to clamp down on the rock. You may never see it move, but it does. Marine snails called limpets are another type of mollusk that scrapes algae and bacteria off the rocks. Some of these form depressions in the rock that exactly fit their shells. They

Green surf anemone (*Anthopleura xanthogrammica*). Up to 11.8 inches (30 centimeters) across.

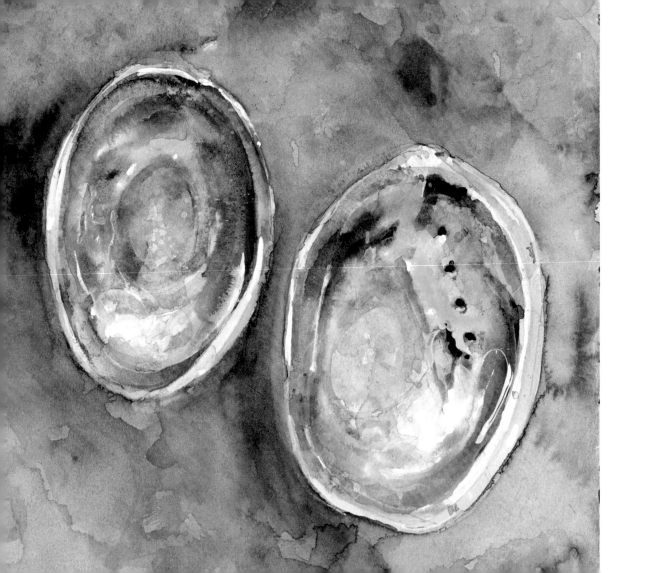

move back to this depression as the tide goes out. They clamp down hard, which makes them very difficult to pry loose and keeps them from drying out in the sun. When the tide is high, covering them with water, they feed. Limpets are snails, but their shells don't form coils. Instead they are roughly cone shaped. Abalones are another gastropod in the intertidal zone. Their shells have holes through which they expel water, waste, and gametes—cells for them to reproduce. Unfortunately, abalone meat is very good to eat, and overcollecting means that there aren't many left.

Crabs with big pincers live under the seaweed or macroalgae that can cover areas of the rocky intertidal zone. Octopuses—whether the giant Pacific octopus that can weigh in air as much as a human or the miniscule octopus that lives under rocks—also live in the rocky intertidal zone. Sea stars are a serious threat to animals who can't run, and they abound in the North Pacific. Recently, warming water has allowed disease to spread and their populations to decline. The impacts of their decline are to be determined. You might think that with fewer predators, life would be easier, but life is never easy in the intertidal zone. Overcrowding can be as difficult to cope with as the threat of death by sea star.

ARTIST NOTE

While painting species from our rocky shores, I couldn't help but feel the presence of the art scene in NYC in the 1940s when abstract expressionism was making a new wave. Rothko's 1944 painting Slow Swirl at the Edge of the Sea *is an example of a vibrant, veiled, obscure, translucent landscape. The painting is more literal than an abalone but they seem related in spirit. Who knew that abstract expressionism could be found all through nature!*

Opposite: Abalone (*Haliotis* sp.). Up to 12 inches (30 centimeters) long.

CHAPTER 3

Gulf of California

My first experience researching the sea was in the Gulf of California, which separates the Baja California Peninsula from mainland Mexico. Its southern tip is entirely in the tropics, and occasionally Indo-West Pacific fish like the Moorish idol show up here, a rarity in North American waters. In the northern end of the Gulf is the mouth of the once mighty Colorado River. Over eons, the Colorado River ate into the bedrock of North America; it carried away silt and sand, depositing them at its delta. The Gulf lies on an earthquake fault, its southern extent a very deep basin with hydrothermal vents. Its depth shallows to the north.

The long U-shape of the Gulf amplifies the tides that enter from the south. The tides get more extreme as they move north; in the shallow north, the highest high tide is 16 feet (4.9 meters) higher than the lowest low tide. The land that is exposed between the highest high tide and the lowest low is the intertidal zone.

This extreme tidal range is hard on animals who live between the tides, but what makes it worse here is the yearly temperature range. The northern Gulf is surrounded by the Sonoran Desert, where high temperatures are routinely over 100 degrees Fahrenheit (38 degrees Celsius) in summer. Its shallow depths and the extreme summer heat warm the Gulf waters by August to nearly 88 degrees Fahrenheit (31 degrees Celsius). In winter the shallow northern Gulf cools to about 62 degrees Fahrenheit (17 degrees Celsius), comparable to Southern California, where no one goes swimming and surfers wear wet suits. Few bodies of water change temperature so much in so little time.

The animals of the northern Gulf adapt; they have to. Sometimes they move deeper, where the temperature is more moderate, but during extreme tides, plants and animals trapped in the intertidal zone have to cope with the outrageous temperatures, often for hours. Some, like the algae that grow on rocks, can't move, and they can't take the heat. They grow when they can. In winter, cold-water algae thrive but disappear without an apparent trace as the water warms. They mysteriously reappear in fall.

I once was out on a massive intertidal sandflat in the peak of summer for the morning low tide. I was researching a small octopus that uses empty snail and bivalve shells as shelter. A group of gulls was making lots of noise nearby, so I walked over. I found an empty shell sitting in a pool of octopus ink. Apparently an octopus was stranded by the low tide, and its shelter was exposed. It must have grown too hot for the octopus in the air, forcing the octopus to leave its shelter (a very risky move for an octopus). The gulls took notice and had the octopus for breakfast!

My research found that octopus of this species live in the intertidal zone through the year. Because they don't generate or maintain their body temperature, the pace of their lives varies with the temperature of the water they live in. In the coldest part of winter, they are essentially moving in slow motion; they hardly venture out. In summer, most of them move into deeper water, except for a few unlucky ones like the octopus the gulls had for breakfast.

The spectacular Gulf of California has many unique species, including one of the smallest porpoises known, the vaquita. It was only discovered in 1958, but it is on the verge of extinction. It may even be extinct by the time you read this book. The main problem is probably fishing using what are called gill nets—uniformly spaced panels of netting that hang down from a line floating at the surface. The target of these nets is a fish called totoaba. These fish used to reach weights over two hundred pounds (ninety kilograms); they'd spawn once a year in the Colorado River delta. Too many have been taken by the fishery, and the Colorado Delta usually has very little fresh water. People in the southwest United States take water out of the Colorado for their own needs, leaving this fish and other animals that used to thrive in the river's delta high and dry.

One porpoise wouldn't do, but two didn't seem much better. I eventually fell into the three-five-seven compositional rule. That rule says it sometimes seems harder to develop a composition with two, four, or six subjects. When you have an odd number of subjects in your composition, it can be easier to move around the page and balance all the parts. And I use the word rule *with hesitation. There really are no rules, just some possibilities that might get you out of trouble. So I added a distant specimen, and that seemed to make a more interesting story. Luckily the heavy paper I used allowed me to use over thirty layers of color.*

I never took a watercolor class, so my method is just an adaptation of oil painting techniques. With oil painting, you can paint over areas you don't like. You can build up to what you want, and you don't need to be sure in the beginning. I initially tried very light or thin paper, typically used for watercolors, and it rippled when I added paint. I found this unacceptable and began buying heavier paper.

CHAPTER 4

Northeast Pacific

The waters off the coasts of California, Oregon, and British Columbia call. This is a special area especially in contrast to the Sonoran Desert; evergreen-clad hills run to the shoreline, and perfectly symmetrical snow-covered volcanoes tower over the horizon. Given the breathtaking splendor of the shore, you might decide that underwater life just can't measure up, but you'd be wrong. In a canoe or kayak, you drift over the cold, clear water and see life teeming in its transparent depths. You won't see reef-building corals, to be sure, but you'll encounter life characteristic of this area and nowhere else.

The most notable North Pacific species to me is the giant Pacific octopus, *Enteroctopus dofleini*. The species ranges continuously from around Japan to off Southern California (at greater depths where it is cooler). Over their four- to five-year life span, this octopus species grows from hatchlings that are about as big

Opposite: Giant Pacific octopus (*Enteroctopus dofleini*). Total length 20 feet (6.1 meters).

as a grain of rice to adults weighing around 110 pounds (50 kilograms) or even bigger. They kill and eat other animals, like other cephalopods (such as octopuses, squids, and cuttlefishes) do, and sometimes even grab birds.

The birds of the North Pacific are phenomenal. Auks, puffins, guillemots, plus the usual gulls and occasionally, far offshore, albatross. I saw my first albatross off the Haida Gwaii (then called the Queen Charlotte Islands), west of central British Columbia. That sighting would have made any day memorable, but the fact that they were soaring over a small pod of very large sperm whales made that day unforgettable.

The productivity of these waters allows us to emphasize these air-breathing vertebrates, the birds and whales, but the animals without back bones who don't breathe air are every bit as amazing. The sea stars of the North Pacific come in a wide range of species and colors. Predatory snails of the Buccinidae are spectacularly diverse in this area. Even the oceanic squids are unique here. One family, the Gonatidae, has one species in Antarctic waters, two in the high Arctic and Atlantic, and sixteen in the North Pacific.

These squids, as do nearly all soft-bodied cephalopods, produce only one clutch of eggs in their lives. Females of this squid family cradle their eggs in their arms until they hatch. Doing so makes them strikingly different from most squid females, who, once they produce their eggs, abandon them, no doubt hoping for the best.

The eight species of salmon in the world include one in the Atlantic and seven in the North Pacific. In the North Pacific, two live off Asia, and five live off North America. These fishes, like the squids, only produce one clutch of eggs, but they don't keep them in the ocean; the adult fishes swim up the freshwater stream in which they themselves developed from eggs to spawn. Then they die.

The exuberance of this strategy seems to reflect an implicit trust that the eggs of both squids and salmon will produce viable young that will mature to represent these adults in the next generation, come what may. They have no "do overs." Scientists don't yet know how accelerating ocean warming will affect them. Individuals of the giant Pacific octopus are increasingly rare in shallow water, where the temperatures are warmest.

ARTIST NOTE

I must admit that I found painting a live octopus challenging. First of all, I was looking through a YouTube screen and deep water at a moving subject. But my years of developing the slow layering technique provided the solution. When you are "slow layering," you begin with a light wash, a teardrop of paint, and a tablespoon of water. Letting each layer dry is essential. If you put down the next layer too soon, you will "lift up" or remove the previous color. The more I layered (sometimes one hundred layers), the more the area developed depth. For example, one layer could be the texture of the octopus and the next its surface color. There is no award for speed. It isn't a better painting because you did it quickly. All that matters is the end result . . . so try it slow.

CHAPTER 5

Kelp Forest

To the west of continents at temperate latitudes, ocean currents and winds combine to pull water from the depths to the shore. When this cold, nutrient-rich water reaches the surface, it supports a great diversity of life. Off western North America, these nutrients act as fertilizer for an underwater forest of giant kelp, *Macrocystis pyrifera*. This alga extends from the seafloor to the surface and can grow as rapidly as two feet (sixty centimeters) per day. Most often they grow at a rate of about one foot (thirty centimeters) per day, until they reach lengths of up to 175 feet (53 meters), with their leaflike blades floating on the surface. Their lengths shelter an amazing diversity of animals.

Among the alga are hundreds of species of invertebrates and fishes, most notably perhaps the golden-orange Garibaldi, the state marine fish of California. Growing on the blades' surfaces are other animals, ones that settle as microscopic larvae never again to

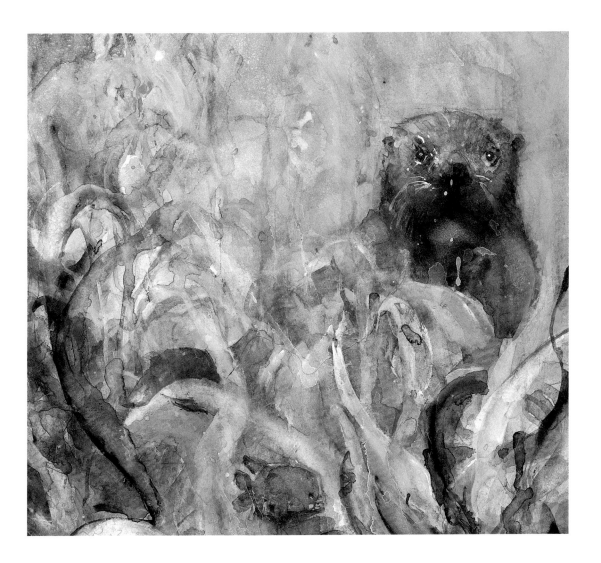

move, only to grow. Like numerous other animals, they live by filtering the seawater with tiny sieves that catch tiny bits of food in the water column. Sponges and tunicates (sometimes called "sea squirts") grow on seafloor rocks, where they provide splashes of color. Bivalves, or clams, draw water through their gills, filtering out nutritious bits that they eat. Sea otters, the fuzzy mascot of kelp forests, in turn eat these clams.

Sustained by cold, nutrient-rich waters, the kelp provides the skeleton of the ecosystem. The kelp also falls prey to killers. Sea urchins, echinoderms coated with spines and armed with mouths made of grinding plates, can form marauding armies on the seafloor. Checks and balances sustain the ecosystems; other echinoderms, sea stars, prey on small urchins.

Giant kelp (*Macrocystis pyrifera*). Up to 175 feet (53 meters) long (*left*). Sea otter (*Enhydra lutris*). Up to 4 feet (1.2 meters) long; 50 to 70 pounds (23 to 32 kilograms) (*center right*). Small Garibaldi (*Hypsypops rubicundus*). Up to 15 inches (38 centimeters) long (*center*). Sunflower star (*Pycnopodia helianthoides*). Typically 15–25 inches (40–65 centimeters) across (*lower left*). Purple sea urchin (*Strongylocentrotus purpuratus*). At base 2–4 inches (5–10 centimeters) wide by 1.6 inches (4 centimeters) tall (*lower*). Bat star (*Patiria miniata*). Up to 8 inches (20 centimeters) across (*extreme lower right*).

This balance is fragile. Warming waters have brought a mysterious disease that kills sea stars. The urchins, now unchecked, form dense armies and make the kelp vulnerable to their grazing. The rich life typical of mature kelp forests may have nowhere left to go if the urchins keep eating.

Top: Giant kelp (*Macrocystis pyrifera*). Up to 175 feet (53 meters) long (*upper left*). Garibaldi (*Hypsypops rubicundus*). Up to 15 inches (38 centimeters) long (*lower left*). Leopard shark (*Triakis semifasciata*). 4–7 feet (1.2–2.1 meters) long (*upper right*). Kelp bass (*Paralabrax clathratus*). Up to 28 inches (72 centimeters) (*lower right*). Purple sea urchin (*Strongylocentrotus purpuratus*). At base 2–4 inches (5–10 centimeters) wide by 1.6 inches (4 centimeters) tall (*center bottom*). Wrasse (*background*).

Bottom: Sunflower star (*Pycnopodia helianthoides*). Typically 15–25 inches (40–65 centimeters) across (*left*). Purple sea urchin (*Strongylocentrotus purpuratus*). At base 2–4 inches (5–10 centimeters) wide by 1.6 inches (4 centimeters) tall (*center*). Bat star (*Patiria miniata*). Up to 8 inches (20 centimeters) across (*lower right*).

The task of painting water surrounding and churning plant life presents quite a challenge. One system that works (ironically) is called the pouring technique. I fill small jars with an inch or so of transparent watercolor and water and pour the contents over a dry surface that is already complete with images. Two or three colors can be used to soften the image and create "sway." I am glad to turn over my process to chance after tediously rendering plant life for days. My base image always survives, but the softening and looseness of the washes create movement inherent in underwater subjects.

CHAPTER 6

Coral Reefs

Not all reefs are coral reefs. A reef is any hard, submarine ridge that comes close to the water surface or even emerges from it. It's a navigational term. Animals such as oysters and tunicates can cover reefs, but reefs made up of coral are the most famous, and deservedly so. Coral reefs are home to a large proportion of all known marine species. These shallow, warmwater bonanzas of life are actually built by corals, a type of animal that needs help to succeed.

Corals get energy by filter feeding—extending their tentacles into the water and snagging particles of food that they bring to their mouths—but where they live, there isn't enough food in the water column to sustain them. They get around that problem by hosting live microscopic algae inside their bodies. The algae, being plants, harvest the energy in sunlight to make sugars, a process known as photosynthesis. A coral gets more energy from these algae than it does

by capturing its own prey. This relationship is called a symbiosis, meaning that living so closely together benefits both parties. The coral gets food. The algae get a safe, sunny place to live and nutrients in an area where nutrients are scarce. To keep the algae in the sunshine, the coral produces a calcium carbonate skeleton that rises above the shadows. The skeletons of all the corals form the reef. The algae photosynthesize and share the bounty gathered from sunlight with their animal hosts. At night the coral extends its tentacles to snag tiny animals from the surrounding water. The tiny animals suitable for coral to eat are scarce in tropical areas where corals flourish, but that food is an important source of nitrogen, a fertilizer for the algae.

This simple animal-plant relationship produces phenomenal ecosystems that are so large they can be seen from space. The individual coral animals or polyps that build this massive reef system are small, even tiny. We may call a massive block of calcium carbonate—the substance that makes up the reef—a coral, but in truth the real coral is an individual polyp, less than a tenth of an inch (two or three millimeters) across. It's clear that the massive blocks result from an incredible number of polyps working together. Because these polyps depend on sunlight to sustain their algae, they grow toward the sun. Over time, they generate a complex, three-dimensional habitat that offers abundant protected nooks and crannies for associated mobile animals. Where corals colonize, over eons, their skeletons build up and moderate any wave action: outside the reef is a zone of high wave activity; within the reef the water is calmer. The more diverse the habitat types available, the more diverse the animals living in them are.

Although many animals use the refuge that the corals create, some are more directly dependent on coral. Parrotfish use their huge front teeth to bite off chunks of coral to eat. A type of clam bores into the coral skeleton, where it resides while filter feeding

Opposite: Southern scythe butterflyfish (*Prognathodes carlhubbsi*). Up to 5.1 inches (13 centimeters) long (*left*). Stylized blue angelfish. 8–14 inches (20–36 centimeters) long (*right*). Rainbow wrasse (*Thalassoma lucasanum*). Up to 4 inches (10 centimeters) long (*lower*).

for the rest of its life. (It releases either eggs or sperm into the water, hoping they meet up with either sperm or eggs released at about the same time by a nearby member of the same species to make more clams.) Butterfly fishes use their long snouts to pick animals off the coral, but they mostly eat bits of the coral itself.

The coral reef system and all its component parts rely on corals being healthy. These animals, however, face many threats, from physical damage to being buried by dirt or soil that washes into the ocean from land (called siltation) to overgrowth by algae to rising sea temperatures. Anything that covers the tops of corals limits their algae's access to sunlight, which they need to gather energy. Warming waters perturb the symbiotic relationship and stress the corals. After a certain point, the stressed corals evict their algae. Because the algae give the corals their colors, after evicting their algae, the corals look white, or bleached. At first this is temporary, and they are able to welcome the algae back when the temperature, and thus their stress levels, are lower. Longer or more severe warming makes the eviction permanent and means the death of the coral. Dead reefs are overgrown by other algae that cover the coral like a blanket. This reduces the diversity of the space and of the animals in the area.

ARTIST NOTE

I find the coral reefs amazing and magical—and their disappearance especially depressing. They don't seem real in the first place, but the visuals created when they are overgrown by microalgae, which brackets the reefs in a white haze, tells the story of their death without words. I first created the reef forms and then simply continued to wash over them with the paint color titanium white and water until the whole scene was made quiet. I let each layer of white dry much like the warming of the ocean. Layering in watercolor is a slow and repetitive task. If a layer of titanium white is cool to the touch then it is not ready for another layer of water/paint. You can edit ad infinitum, which makes the process one of discovery and persistence. These are qualities greatly needed in addressing our pressing environmental issues today.

Healthy reef (*left*); smothered reef (*right*)

CHAPTER 7

Indo-West Pacific

Most of this book focuses on the East Pacific Ocean, as this is where most of my at-sea experiences have been. However, any book about marine life simply has to give a nod to the Indo-West Pacific, the huge expanse of tropical ocean from eastern Africa to the Pacific islands of Hawaii and Easter Island. Coral diversity is especially notable in the center of the Indo-West Pacific, an area termed the Coral Triangle. Stretching from an apex at the Philippine Islands south to the eastern Solomon Islands and west to Borneo, this triangle is home to over four times as many coral species as is the Western Atlantic. And it's not just that the corals are so diverse; more than half of all the known reef fishes from the entire Indo-West Pacific can be found in the Coral Triangle, which has more reef fishes than anywhere in the world. In fact the Coral Triangle is home to more than twice as many fishes as the Western Atlantic.

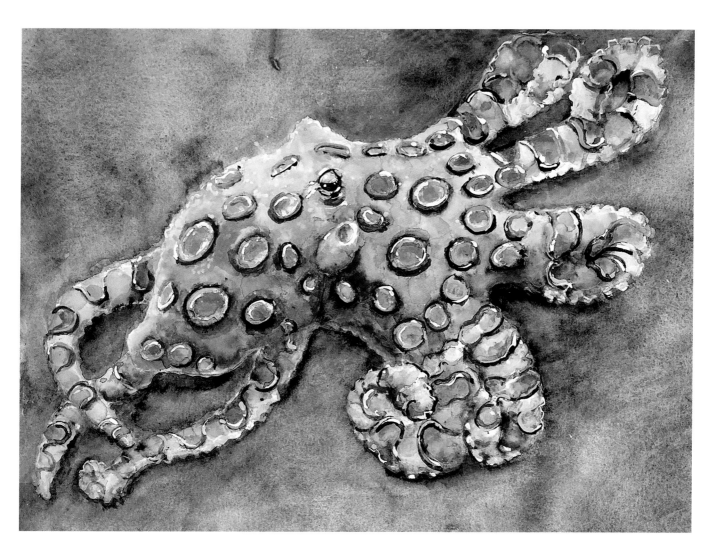

Why aren't we comparing the Indo-West Pacific to the East Pacific? That comparison wouldn't be fair, as the East Pacific has what amounts to only a fraction of the shallow-water area the Indo-West Pacific has. The deep sea comes so close to the shore in the East Pacific that there simply isn't room for much tropical diversity there. In addition, the prevailing winds tend to push the ocean water away from the shore; the deep water (which tends to be cold) rises to fill in the gap. The total area with tropical temperatures is also much smaller in the Eastern Pacific than in the Caribbean, where the edges of continents and islands provide abundant shallow-water habitats where coral reefs can thrive. The East Pacific is nearly bereft of islands, another reason why its warm, shallow-water habitat is so scarce. The warmth of the water doesn't absolutely mean higher diversity, but we'll consider that below.

The Indo-West Pacific is home to many small islands that offer complex habitats. Half a world away from where I write in a chilly Chicago, the Indo-West Pacific seems deliciously tropical; its sea temperatures are warm, nearly room temperature, year-round. And those temperatures are thought to have been mostly constant for a very long time (an estimated thirty-five million years). This history of stability makes the area unique. The Ice Ages that produced continental glaciers as far south as Chicago passed unnoticed in this area centered on the equator. Some scientists think constant environments are key to allowing species, once evolved, to survive for a long time.

We have created a new problem for these long-surviving species. The water in which they live is warm, but now it is getting warmer rapidly. If these species reached such high diversity because their habitat had been stable for tens of millions of years, they may lack the ability to adapt to environmental change. In addition, a lot of people live on the land surrounding this ocean area. Large human populations result in, almost without exception, increased pollution and waste, both of which tend to wind up in the ocean. After all, the oceans are so huge that we think that a little bit of pollution won't hurt much, and when waste is out of sight it is out of mind. The time is coming, in fact it may be here right now, when those ideas simply have to be put to rest.

Sometimes you just have to wash away everything you have painted for days and apply new color. I painted the octopus white with orange and blue spots. But the bright reflective fluorescent coloring just wasn't coming through. Fluorescent coloring refers to colors that absorb more light than common colors. Fluorescence is a form of luminescence—when an object glows without being heated. This makes the fluorescent colors much brighter. Janet suggested I change the white to an orange-ochre color. Orange is the complimentary color to blue; when they are placed next to each other, both colors shine. So I washed away my many layers of white. Yes, I put the piece under the faucet and removed what layers I could. There is always time for reflection and reparations on 300-pound paper. The piece of paper is not literally three hundred pounds. Paper weight refers to the thickness of the sheets and gives the weight of a ream (five hundred pages) of the paper. The heavier the paper, the thicker. The average watercolor paper is much thinner (140 pound), but 300-pound paper will not ripple, and it will allow you to wash off an area, change it, and improve it.

When painting, you just start somewhere with a vague idea and make "corrections" or "adjustments" as you go. It is a process of forgiveness. You don't allow thoughts like "I'm not good." You replace them with "How can I make this better?" Just refine and build, and sometimes you get lucky and the piece works.

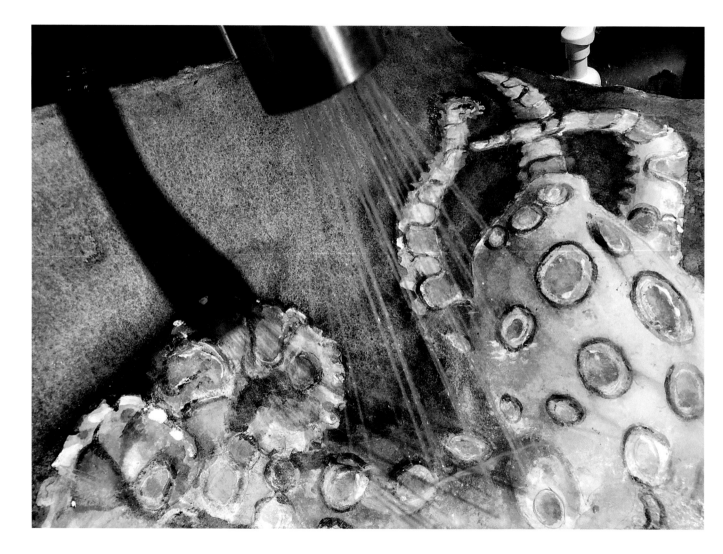

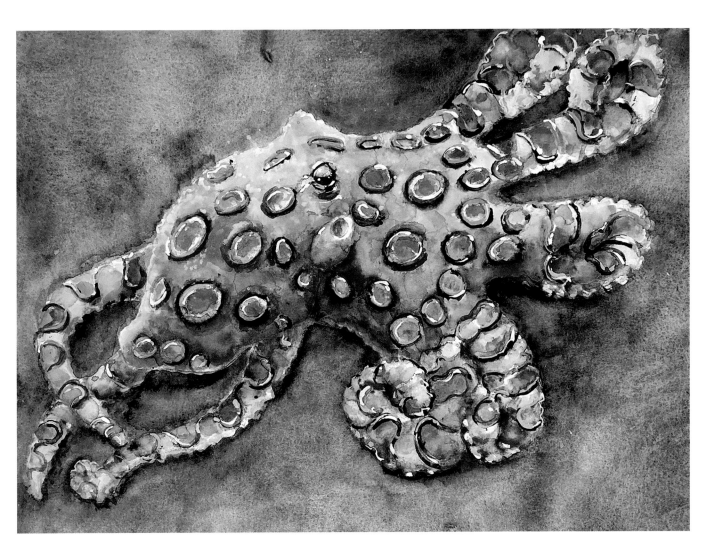

CHAPTER 8

Research Cruise Experience

A lot of scientists spend their entire careers happily studying shallow-water habitats and animals like those discussed above. Some of us want more. We want to see and learn about the ocean away from land. Looking at a map of the world makes you quickly realize the oceans are the biggest habitat on earth. If you then think about the oceans as three dimensional, with a vertical dimension that extends from ankle deep to almost 6.8 miles (11,000 meters) deep, you fur-ther realize they offer the largest area for life to exist on the planet! Sign me up! The only catch is how to get there? I would tend to avoid trying to hitch a ride on a freighter or going around the world solo in a sailing ship, so being part of a cruise dedicated to research is the best option.

Going on a research cruise sounds great. People pay big bucks to go on cruises! The reality, as with so many things, can be a bit different than expected. Rarely do

you meet the ship at a posh passenger terminal but most often at an industrial port. Hulking freighters loom, and rats and the mangiest cats I've ever seen patrol the docks. The ship itself is perhaps better termed a boat; as compared to commercial ships, it's pretty small at shorter than a football field in length. You will be on it with up to twenty-one other members of the science party, which includes everyone aboard the ship except for crew, for up to the next month, maybe longer if you are sailing to a site far away from the nearest port. You may not know anyone else in the science party, but you will in due time.

Shipboard space is limited. Everyone shares a stateroom with someone else and a toilet/shower with two people in another stateroom. The stateroom, despite its grandiose name, has bare walls, a dogged porthole (which in shipboard parlance means it has clamps that secure its cover so even if the glass is broken during a storm, it will not leak) (those staterooms below decks, of course, have no portholes), and indoor/outdoor carpet, if any, that has seen better years. There are two bunks with curtains so that one cabinmate can get up and prepare to stand watch while the other is sleeping. There are also life jackets and survival suits. These would be vital if there were a serious issue with the ship and you faced being on your own in the water until rescuers arrive; you will be trained how to use them.

Ideally, Science Operations (Sci Ops) go on 24-7; most cruises use a four hours on and eight hours off schedule, so you might work from 12 to 4 both a.m. and p.m. That is about the worst watch for me. You miss meals (served buffet style) in the galley as they are only available for about forty-five minutes. Of course, there is food available all the time because people are working all the time, but the prepared meals are usually very good, and it's a shame to miss them. The mess (no one calls it a dining area) doesn't have enough seating for the crew and science party at the same time, so people don't linger over meals in order to let someone else use their seat. There is no wine with dinner; in fact, US research vessels are dry.

I know there is a story behind that, but some things are better left unsaid.

The ship's furnishings are solid and bottom weighted. All closets and cabinets have extra fasteners. Counters in the lab spaces, located in prime real estate on the main deck, are speckled with predrilled holes and have conspicuous rims along the edges. As the science party moves in, the experienced members put placeholders in the lab space as dibs to mark their territory. They open containers of eye bolts and bungee cords and start screwing the bolts into the counters and attaching bungees across their stuff. You might think this is a severe level of dibs, but the first time the ship at sea starts to pitch or roll, you realize why tie-downs are a priority. And why the chairs are so dang heavy. Things that are not secured go flying. Watching someone's reaction when this happens to their stuff is a great way to learn about your shipmates.

As time at sea goes on, you increasingly realize how small the ship is. There is simply nowhere to get away from people but out on deck. You can look out at the sea and think about the wonders it holds and how achieving the goals of the cruise might discover more. Then you're ready to once again join the rest of the science party and work to make those goals a reality.

The abstract art movement was a departure from rendering reality and the accurate depiction of imagery. Total abstraction bears no trace of any reference to anything recognizable. This art movement began in the 1940s in New York City. It first appeared with artists like Russian painter Wassily Kandinsky at the end of the nineteenth century and developed throughout the twentieth century. Now while studying this movement I have never felt any need to be part of it. But when painting water and the creatures that live within it, I feel I have slipped into a form of abstract art. While adhering to the principles of classical representation, my end product seems otherwise. Maybe there is an element of abstraction in all reality. My experience makes me believe abstraction is an essential part of reality. There is nothing plain about reality. If you slow down and observe carefully, there is endless magic.

CHAPTER 9

Open Ocean

As your ship sails off, you realize that nothing compares to the waters of the ocean, far away from land where the land-derived clouds of sediment and abundant algae turn the water a greenish opaque hue. Here in the open ocean, sediment is minimal. Sunlight enters the water, and from a boat's deck about ten feet (three meters) over the sea surface, it looks like it filters down and down forever.

However, that sediment from land carries nutrients; the absence of sediment means low productivity. The open ocean does not have any algae on stalks because the seafloor is much too deep and dark for them to grow. And the single-celled algae that float in the water are also scarce, being nutrient starved. But only these plants can harvest the energy the sunlight gives. Animals, depending on that plant-fixed energy, are also limited in abundance; they are, however, stunning in their diversity.

Fishes, like the tuna and the ocean sunfish, are among the most familiar open ocean animals. These animals grow to large sizes (think in terms of tons) and travel large distances, very quickly in the case of tuna. Ocean sunfish are among the strangest-looking fish in the ocean. They grow to be huge, but they aren't fish shaped. From a research ship, you can see them sometimes floating near the surface. Their common name, sunfish, comes from their habit of sunning themselves by swimming on their sides to get warm before they descend into the depths and look for more jellyfishes and other things to eat.

Jellyfishes, despite their potential to be large, are more likely to drift with the currents than to swim because most of their bodies are water. That doesn't stop jellyfishes from being a favorite prey of sea turtles that range broadly in the oceans. Turtles' appetites for jellyfishes mean, unfortunately, that they will also swallow their look-alikes, plastic bags. Sea turtles breathe air, and sometimes, if you're lucky, you can see them at the surface taking a breath. Sea gulls get tired of flying all the time, too, and they will rest on the back of a turtle on the surface. A turtle that has eaten plastic bags will also come to the surface, but you can tell the turtle is ill and the animal isn't going to get well. The birds don't even sit on it.

The feeling of space and beauty that the ocean gives on a calm day is serene. The power of the ocean far away from land during a storm is terrifying. That's another story.

Opposite: Ocean sunfish or common *mola* (*Mola mola*). Roughly 8 by 6 feet (2.5 by 1.8 meters), but can reach 14 by 10 feet (4.3 by 3 meters) and can weigh over 4,000 pounds (1,814 kilograms).

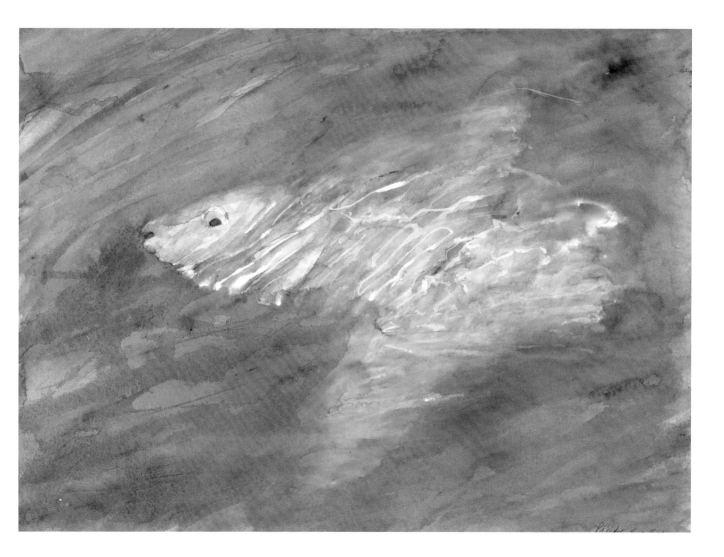

Opposite: School of blue-fin tuna (*Thunnus thynnus*). 6–10 feet (1.8 to 3 meters) long; weighing up to 1,500 pounds (680 kilograms).

Jellyfishes Phylum Cnidaria Class Scyphozoa (200 named species). 1 inch to (rarely) over 6 feet (2.5 centimeter to 1.8 meter) across.

Whereas a drawing is right or wrong, a good composition is debatable. The best quote on the subject is by Polish composer Frédéric Chopin: "When one does a thing, it appears good, otherwise one would not write it. Only later comes reflection, and one discards or accepts the thing. Time is the best censor, and patience a most excellent teacher." I moved the tuna around for weeks until I felt comfortable. There is almost no better way to get it right than to get it wrong first. So don't be afraid to take a hunch and try it. Also, the deep-sea background presented a challenge. I find it easier to wash through the fish (any subject) rather than trying to paint around it. If I paint around it, I get odd edges. Painting through the tuna in this case, makes it one space in time. Later I am able to define the larger closer tuna and give the piece depth. Aerial perspective, representing distant objects as fainter, is a useful tool.

CHAPTER 10

Subsea Vehicles

If you are aboard a research ship, you aren't there to enjoy the salt air. Berths (spots aboard) for the science party (which is everyone who isn't crew) are very limited. Everyone has a job to do. For a deep-sea biologist, the most basic sampling method (simplified a bit here) is trawling. A large net, attached to the ship's winch by a line 2.5–3 miles (4–5 kilometers) long, is thrown in the water as the boat slowly moves ahead. It is allowed to sink and the line is "paid out" for an hour or two or four depending on depth, then the winch stops and the ship drags the net across the sea bottom for a few hours. The winch is then reversed and pulls up the net and its contents. The net is lifted onto the deck with the aid of the science party and its contents examined. Species targeted by the grant that funded the cruise are removed post haste for processing, and the rest of the catch (the bycatch) is up for grabs, but often it simply has to be tossed overboard. The problem is that

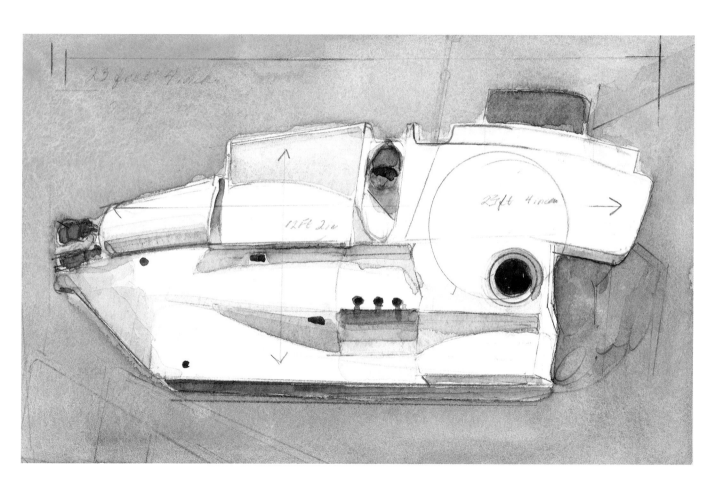

there can be so much of it. Because ship time is so expensive (think ship's fuel, crew wages, food for everyone aboard), the work goes on twenty-four hours a day. Processing a full net of animals, taking tissue samples, preserving the rest of the animals, and labeling both can barely be completed before the next trawl returns.

The net contains a bonanza of life, or what was life. Nets, which have been used for centuries, damage everything, including the seafloor. Although they can deliver rare species for study, they might not. There is no way to target anything for collection. If you do get what you want, you can't control if they arrive mangled or intact. And we can't know their story. Is that rare octopod covered in mud because it lived near the bottom, or did it get muddy in the trawl that bit too deeply into the seafloor? You can't say. For instance, several years ago, scientists working with trawl-collected animals thought that dumbo octopods spent most of their time swimming in the water column, but now we're reconsidering.

Why reconsider? New methods mean new data and insight. Technology is improving how we see deep-sea animals in their natural environment. The first in mainstream use may have been the towed camera. These cameras are set to take photos at given intervals. They are lowered from the ship with a set of very bright lights to what is hoped to be a safe depth, close enough to the seafloor to see something in the photos but far enough above it that it doesn't hit anything. In the early days, of course, the cameras used film, which had to be developed later. At best the film was developed in a darkroom on the ship to verify that the photos showed something and that the depth allowed the seafloor to be imaged. At worst the film was developed on land after the cruise ended. You can imagine the frustration.

To be actually on the seafloor, the only place on the planet where no one has walked, has been a dream. The first subsea vehicle to carry people to the seafloor and make important scientific contributions was the *Alvin*. Owned by the United States Navy and operated by Woods Hole Oceanographic Institution,

Alvin carries three people to the seafloor. It can dive to depths of over 2.7 miles (4,500 meters) and returns to the ship, having completed scientific observations and tasks. The work is actually performed by an experienced pilot inside the vehicle who drives *Alvin* and uses its arms or manipulators to sample, deploy, or recover experiments. The two other people in *Alvin* are termed *observers* (it's a hint that they aren't supposed to touch anything). One usually is an experienced *Alvin* user and sits to the pilot's left; the other is usually more junior and sits to the pilot's right. Because junior observers aren't expected to say much worthwhile, the pilots tend to ignore them.

A sphere seven feet (two meters) in diameter of inch-thick (2.5 centimeter-thick) titanium forms *Alvin*'s personnel carrier. It is minimal space for three people, oxygen tanks, carbon dioxide scrubbers, fire extinguishers, controls, computers, and video recorders that immortalize what is seen in front of the sub. Observers operate the outside video cameras via pan and tilt controls and zoom and focus.

Under the faux floor that provides a level place to sit are emergency supplies. *Alvin* has no insulation to speak of. The ocean, below depths of roughly 3,200 feet (1,000 meters), is a fairly uniform 35 degrees Fahrenheit (2 degrees Celsius), about the same as the inside of a refrigerator. It gets cold. If *Alvin* should be immobilized—for example, if it were to snag on fishing gear abandoned on the bottom—sleeping bags stowed under the false floor would help keep the divers warm enough to survive till they could be rescued. There is enough oxygen to keep three people alive for seventy-two hours.

Outside the sphere, *Alvin*'s white fiberglass covers shield the batteries that power everything *Alvin* does and the syntactic foam that provides buoyancy. Thrusters mounted aft, on the back of the vehicle, provide propulsion (at a slow walking speed). Forward is "the basket," where we secure boxes in which we place material collected from the seafloor to keep it protected during ascent. View ports are the windows. They are about as big as your face, and through these you look out into the ocean depths. *Alvin* began work in 1964 and remains current by undergoing upgrades every three years. In 2011–2013, a major upgrade expanded its sphere and modernized its controls. Its mother ship, the RV *Atlantis*, has dedicated space for *Alvin*. Given that *Alvin* weighs 45,000 pounds (20.4 metric tons), it is not about to be shoved aside.

More recent technological advances in deep-sea research are most notably Remotely Operated Vehicles (ROVs). They are called remote as the pilots stay aboard the ship with the entire science party and send commands to the vehicle via fiber-optic cables. The vehicle livestreams video from multiple cameras through the same fiber-optic cables, allowing the pilots to see so they can drive the vehicle, sample, and make deployments and recoveries. Recent innovations mean ROVs can livestream what they are seeing to the internet, and you can watch in real time as long as you're connected.

My first view of the seafloor came via the eyes of the ROV ROPOS (Remotely Operated Platform for Ocean Science). I was enthralled. Even better was the chance to see it with experienced members of the science party, including both geologists and other biologists. I learned so much! Unlike in *Alvin*, when I wanted a cup of coffee, I could go to the galley and get one; the bathroom (the head) was just down the passageway. In their early days, putting a robotic vehicle with complex electronics and hydraulic systems into the ocean and lowering it to depths of 1.5 miles (2,400 meters), where the hydrostatic pressures exceed 3,494 pounds per square inch or 237 atmospheres (1 atmosphere is the air pressure at sea level), was asking for a breakdown. This made (I think) getting on cruises easier. Broken hydraulic lines or short circuits force the vehicle to return to the surface. The technology, however, is advancing all the time.

There are now Autonomous Underwater Vehicles that are programmed to go off on their own and survey a part of the seafloor, sensing temperature anomalies or other such oddities.

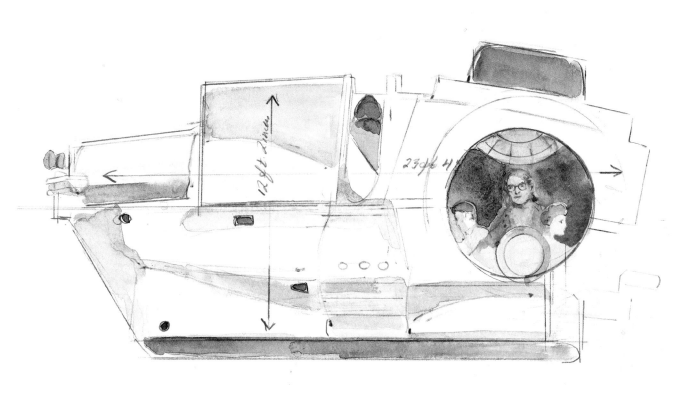

Alvin cutaway showing Janet and other passengers.

Diving in *Alvin* is an exhilarating, miserable experience. You have to enter the personnel sphere from the top, fourteen feet (four meters) high. You go up a ladder attached to the A-frame (a hoist descriptively named) and cross a catwalk that shakes under you (or maybe it doesn't, but I just remember it that way, being afraid of heights). You take your shoes off, as you won't need them where you are going, climb into the orange sail on top of *Alvin*, and descend via a ladder into *Alvin*. You try to get comfy as you will be there for the next eight hours (hopefully not for the next seventy-two). After everyone is in, someone lifts the ladder out of *Alvin*, closes the hatch in the orange sail, and bolts it from the outside. Given the immense pressure at depth, it seems that bolting it closed is overkill, but they follow procedure. If you are going to be claustrophobic, now is the time.

By this point, you will have had a predive period inside *Alvin* with a pilot and another observer. The pilot spends about thirty minutes showing you things in the sphere to make sure you're ready to make the most of the dive time. In fact, I suspect the pilot is watching to see if anyone seems nervous. Another part of the predive training is to make sure you can don an oxygen mask and have it seal tightly enough to your face that there are no leaks. If there should be a fire inside *Alvin*, you will have to put this on, and it can't leak. The pilot would put out the fire by releasing inert gases in the sphere to reduce the oxygen levels so much that the fire couldn't burn, and of course that you couldn't survive without that oxygen mask.

These things go through the first-time observer's mind as the pilot is on a radio checking with experts aboard the ship to verify everything is prepared for the dive, your dive. Suddenly, the ship's motion to which you've grown accustomed over the previous days or even weeks changes as *Alvin* is lifted off the deck by the A-frame. When the A-frame swings aft and lowers *Alvin* into the water, you feel the waves again. There suddenly are people in the water, swimming around *Alvin*, making sure everything in the basket is secure and whatnot. When they give the OK, the pilot fills the ballast tanks and *Alvin* sinks.

On my first dive, the thought that kept coming to mind was that I was off the ship in the North Pacific in a tiny vessel. How could we ever be found? This was silly of course; there are radios and sonar beacons, and the ship knows where *Alvin* is every second, but after you start to realize how big the ocean is, it takes on a new reality. As I looked through the viewport, the sunlight was fading at eight o'clock in the morning, and everything felt alien.

As it got darker, and bioluminescence sparkled, my mood eased. My colleagues and I reviewed the dive plan, what we were to do, in what order based on priority, and I kept looking out the viewport. Time came to put on the extra clothes we had packed the night before to stay warm. No fleeces are allowed; when polyester burns, it releases a toxic gas, so only cotton and wool are allowed in the sphere. The cold was starting. You don't see your breath in *Alvin*, but the moisture you exhale condenses on the coldest surfaces. At first that's the viewport, where water pools at its lower edge. A leak? It's a scary thought. Over the course of the dive, water condenses on the walls and dampens the pad on the faux floor. Your knees and back start to ache, and you'd like to tell the guy across from you to move his foot, but it's kind of warm so you hold off saying anything. Suddenly, you're in the deep scattering layer and nothing else matters. Then you're in the dark again and the bottom is in view. You have work to do. *Alvin* has minimal communication with the ship and the scientists aboard it, so the observers in the sub have to optimize what to do based on the dive plan and what they see from the viewports. Failure to optimize wastes the precious battery power and shortens your dive and can leave important objectives uncompleted. The chief scientist has to have confidence that the observers will do what she would do if she were in the sphere herself. Overruling the dive plan is sometimes the thing to do but risks censure. If you overrule the dive plan and get everything done, no one pays any attention, but if in doing so you miss something, you can be rebuked and threatened with not being allowed to dive again during that cruise.

The whole process is not easy on anyone. The observers have physical discomfort (and a notable lack

of bathroom facilities); the scientists aboard the ship above and leading the cruise have to trust the observers and the clarity of the dive plan fully. They can only hope that *Alvin* reaches the bottom in the right place and that what the observers see when they get to the bottom is what the dive plan says.

As chief scientist, I always tried to take a nap while *Alvin* was diving, after lunch, of course; there is precious little I could do anyway. And while *Alvin* only dives during daylight for safety, Science Ops continue twenty-four hours a day to maximize benefit from having the ship on site.

The experience for the divers is worth every bit of discomfort, cold, clamminess, and doing without that morning cup of coffee, as they get to see what no one else in the world has ever seen and live to tell the tale.

Opposite: Giant Pacific octopus (*Enteroctopus dofleini*) (model). Total length 20 feet (6.1 meters) (*top center*). Three common octopus (*Octopus vulgaris*) in jars. Total length up to 4.3 feet (1.3 meters). Vent octopus (*Muusoctopus hydrothermalis*). Total length 7.2 inches (18.4 cm) (*bottom left*).

ARTIST NOTE

The artist searching for new methods, fresh subject matter, and unique presentations is likely to attempt adventures similar to that found diving in the subsea vehicle. But the artist, although also heading into unknown territory, is in no danger and seems no more than a curious soul seeking new visuals. Janet on the other hand is going where most of us never will.

Seeking the unknown is part of the artistic process. I think it is important to note that you master techniques that went before and then continually push toward finding new ways to express your ideas. I think if you have an inkling that a new medium, combination of images, or form of expression might be worth trying . . . then just go ahead. My "Behind the Scenes" series (like the piece above, "Octopuses in the Field Museum Collections" watercolor, 22 by 30 inches/56 by 76 centimeters) came naturally. It doesn't need to be forced. The history of art through the ages reflects the simple truth that there is always more to be discovered and that ways of presentation are endless.

CHAPTER 11

Gulf of Mexico

I will make one more exception, the Gulf of Mexico, to our focus on the eastern Pacific. I've been on several cruises to the Gulf of Mexico, and it seems biased to omit it. This Gulf lies south of the continental United States and east of Mexico. Near the center of its northern coast lies the delta of the Mississippi River, the tenth largest river on the planet. It carries sediment from thirty-one states and two provinces in Canada. All that sediment is discharged into the Gulf of Mexico, where it covers its northern extent.

The Gulf wasn't always like this. It probably didn't even exist before the age of the dinosaurs, and after it formed, a salt layer was deposited in its northern part. Salt is odd to see under water, and it suggests that the sea dried up at some point. Regardless, that salt is important as it traps petroleum—petroleum and shrimp are what the northern Gulf of Mexico is famous for. Sailing out of a port in southern Louisiana, the ship

LOUISIANA

goes through lanes bounded by oil and natural gas platforms. Hundreds of them. These platforms are connected by pipes to oil resources under the surface and probably under the salt layer that caps the flow of the oil and gas. Methane (natural gas) can't be avoided when collecting oil. It costs too much to keep it around (and it is highly flammable and toxic), so the people running the platforms burn it, at night. Huge flames erupt out of the tops of the platforms. It looks like something catastrophic has happened, but no, it's just economics controlling by-product disposal.

That gas and oil isn't just sitting sealed under the salt layer. It's been leaking out for ages. The evidence: the Gulf has natural oil slicks that appear and disappear. When deep-sea researchers took subsea vehicles to the bottom, they found strange assemblages of animals, notably tube worms and mussels that appear similar to those that had been recently discovered at hydrothermal vents, but there was no heat. These animals are all dependent on bacteria that convert the petroleum products into energy they, as animals, can

use. In contrast to hydrothermal vents, where life is short and the environment chaotic, life at cold seeps is in the slow lane. Tube worms here are estimated to live for centuries, up to three hundred years.

The seeping petrochemicals alter the environment. Through chemical reactions guided by bacteria, limestone is formed, a molecule at a time. This rock, elsewhere rare in the sediment-covered northern Gulf, offers homes to larvae of corals and other animals that can't abide so much dirt!

Studies of the seafloor from subsea vehicles also discovered frozen natural gas, called methane hydrates. It isn't that it is so cold that things ordinarily freeze, but methane does due to the combined cold and pressure from the overlying water. Hydrates form large white structures, and as the vehicle bumps them, you can watch bubbles of natural gas form and escape. That frozen methane, when collected in big enough chunks, can be brought aboard ship and lit. It burns with the blue flame characteristic of pure natural gas.

You may remember the huge (130 million gallon

[492 million liter]) oil spill that happened over a decade ago in the Gulf of Mexico. Eleven people died. Something like that will happen again. Industrial accidents are inevitable. The impact when they happen at sea remains to be fully understood. To try to minimize the effect on the surface waters, and those economically important shrimp, chemicals intended to break up the oil, called dispersants, were released at the well head; the oil flowed out unabated and was followed by barrels and barrels of dispersants for a very long time. After the well was capped and the oil and dispersants were no longer released, scientists in subsea vehicles documented that the oil and dispersants caused problems for the deep-sea animals seven years later. The animals affected range from tiny forams, which are food for many of the smallest animals, to corals that are hundreds of years old.

The week I teach perspective drawing is always a challenge. I'm not quite sure why. It's one of the gifts of the drawing process. You lay out your subject, then guess where the eye level (horizon line) would be and draw it across your subject. Then you measure one angle (a wall is higher at one point and lower a short distance away). Then you drag this angle down until it reaches your horizon line and you have a vanishing point. Then you bring all other parallel angles to this point and your piece looks like you know what you are doing. If you are facing a flat wall, it will be one-point perspective, like disappearing railroad tracks. But when you are facing a corner, as with the oil rig, the lines are heading in opposite directions, and you need two vanishing points. This is two-point perspective. The right side of the oil rig is leaning down and to the right. I took one level and continued this line until it hit the horizon line. Where it hit was the right-side vanishing point, and all other parallel lines on the right went to this point. I did the same on the left side of the oil rig. Both vanishing points are off the paper as you can see. Try putting some tracing paper over my painting and continue the lines of the rig and you will see how two-point perspective works. I know it is hard to grasp with language, so try it a couple of times and surprise yourself.

CHAPTER 12

Midwater Depths

Beneath the sunlit waters of the open ocean—after all sunlight is reflected or absorbed by specks in the water column—we enter the largest habitat on Earth: the midwater depths. Defined only by what it is not, its name says it is not the sunlit shallows and not seafloor. But this area, despite its lack of the sun's energy, contains the least known ocean animals, animals like the colonial siphonophore, a distant relative of corals and anemones, and dark red jellyfishes and ctenophores so fragile they fall apart when handled. Their home is the largest space on Earth. They live on food that falls from above: feces, mucus, and bits a predator drops while devouring its prey. Life is hard, as good food is scarce, and every animal tries to get more.

Camouflage would seem to be useless as there isn't, even at high noon, any sunlight. The bodies of animals who spend their lives at midwater depths are often clear or red (except for the eyes and guts, which can't be made clear and still function). As the function of the eyes is to capture light, they couldn't do that

if they were transparent and let light flow through them. If the guts were transparent, when the animal ate something that was opaque, the gut would become visible. This would be an unacceptable risk. The eyes and guts instead are often shiny, being covered with reflective tissue. Red is a frequent color because red light is absorbed by water very quickly, meaning that red looks black from just a few feet away. Animals here frequently have light-generating organs that contain luciferin, which with an enzyme emits light. In some animals the light-generating cells are covered with a lens that focuses the light or with shields that cover their light when necessary. An estimated 80 percent of animals at midwater depths are bioluminescent; they glow because they make their own light. These lights may attract other animals, such as potential prey or sex partners. But they also attract potential predators. If the animals sense a predator is nearby, they douse, or cover, their light, and by moving only a few body lengths in any direction, they are likely to escape. Without lights for signaling through the clear mid-water depths, the animals would have to rely entirely on chance to meet other animals in their immense three-dimensional habitat.

What kinds of animals live in this habitat? All sorts! There are fishes, like the angler fish and gulper eel, mollusks like squids and special swimming snails, crustaceans like amphipods and isopods, and cnidarians like siphonophores and jellyfishes that can grow to huge sizes.

The species of these groups that live in midwater are able to float effortlessly, buoyed up by low-density fluids. Fishes have long tails, big eyes, and very sharp teeth. They also have huge mouths and stomachs because food, when and if it comes, may be in a large package. Big mouths and stomachs mean they can eat prey nearly as big as they are. How long do they take to digest such a meal? We don't know, just like so many things in this part of the ocean.

*Opposite: Gulper eel or Pelican eel (*Eurypharynx pelecanoides*). 2.5 feet (0.75 meter) (*right*). Siphonophore Phylum Cnidaria Class Hydrozoa Order Siphonophorae (very little known; over 175 species named). Up to 130 feet (nearly 40 meters) long (*left*).*

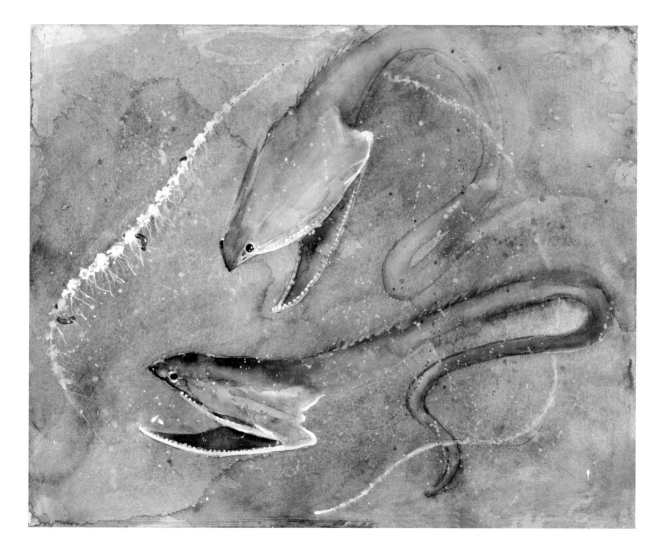

At the midwater depths, things really get crazy. The angler fish and the gulper eel were new to me. These creatures can be likened to the characters in a graphic novel. Again nature foreshadows an art movement. Historian Richard Kyle coined the term graphic novel *in an essay in 1964. This art form has grown in popularity and brings new meaning to the term* expressionist. *With the graphic novel, images became "fantastic and expressive." Similarly, the creatures like the angler fish seem "made up." Who knew there were superheroes in our oceans!*

CHAPTER 13

Deep Scattering Layer

Midwater depths are not uniform. The first sonar reflections during World War II showed the existence of a layer that reflected sonar (sound) pulses far above the seafloor, indicating that something was closer to the surface. As these uniform, sunlight-deprived depths were thought to have very few animals, this was somewhere between a surprise and a mystery. The discovery these images revealed was a concentration of animals that make the greatest migration on Earth. You might think of migrations as the journeys birds make twice a year, moving between north and south with the seasons. Animals in what's called the deep scattering layer, because it scatters sonar reflections, move from deep to shallow with nightfall. Most of the animals involved are small (they might readily fit into the palm of your hand), but they move 2,200 to 3,000 feet (700 to 900 meters) or more up and down every night. That's a long way to go for a small animal.

Driving this migration is the need for food and the fear of predators. Midwater depths don't have a

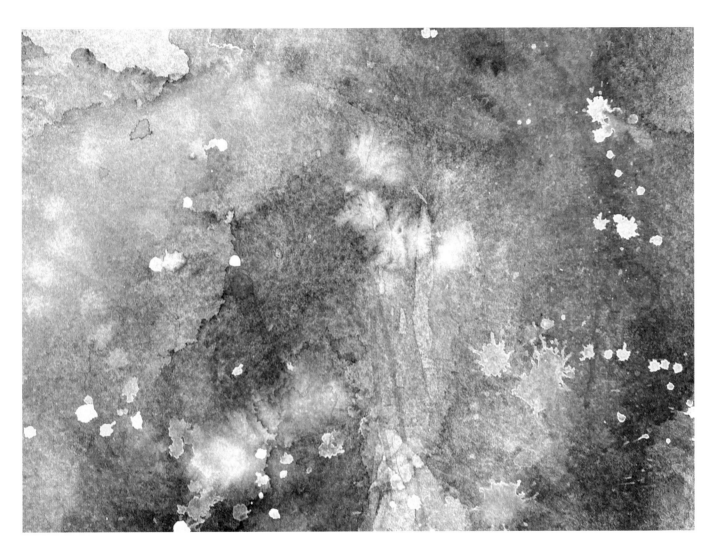

lot of food. Sunlight supports photosynthesis in algae in shallow depths, where animals that eat algae hang out. At those sunlit depths, however, are also predators of small, palm-sized animals who hunt by sight. Animals in the deep scattering layer have the best of both worlds. They live in the dark, food-poor areas during the day when predators patrol sunlit waters looking for someone to eat. As the sun sets and darkness limits predators who hunt by sight, these deep scattering layer animals begin to move into those upper waters with their relatively abundant food. Deep scattering animals move by the thousands. They may not all move all the way up to the surface, but some do. Others stay in the upper few hundred feet where the influence of sunlight is reflected in the increased abundance of animals. They eat; they feast! And with full bellies they sink as the sun nears the eastern horizon. The most common type of fish in the world, myctophids or lantern fishes, are numerous among the animals undertaking this migration, but so are the enoploteuthid squids. In both these fishes and squids, their dazzling light-generating "photophores" form distinct arrays on their bodies. These are how you tell the species apart once you have the animal in your hand.

Peeking out from inside a submersible drifting down from the surface, you notice that the ocean is slowly getting dark. Your first view of bioluminescence might come about 820 feet (250 meters) down—although the water isn't completely inky yet—just isolated points of blue-green light. As you continue to sink, sunlight dims to twilight, which gives the feeling that the light is all around but without a source. Then suddenly you notice that inky blackness has engulfed the sub.

Flashes of bioluminescence become more common as the sub sinks farther, agitating the water column with its turbulence. You press your face against the viewport to see what makes the individual specks

Opposite: Lanternfish Family Myctophidae (246 named species). About 1 to 6 inches (2.5 to 15 centimeters) long, with most averaging under 6 inches (15 centimeters) long.

of light. Suddenly, they are everywhere, in myriad shapes. You flash the sub's outside lights—just for a second—and outside every invisible animal within the range of that light flashes back at you, as if to say, "What are you?" It's like you're in the middle of the Milky Way, only the closest stars are near enough to touch. Then as you continue to sink, the flashing wanes. Mostly the lights have gone out. As the sub sinks farther, the lights become scarcer, then they are nearly gone. You've just gone through the deep scattering layer and you disturbed all those animals resting in the dark with full bellies.

When your submersible leaves the seafloor, you'll pass through the layer again. But because the sub's turbulence won't disturb the animals as much as it ascends, you won't see as many of them. You just made a visit to a habitat that no one knew existed even eighty years ago.

ARTIST NOTE

Painting what I haven't seen has its challenges. I play the underwater video and take still shots with my phone. The task will be getting the glowing tiny fish to appear against the dark spotted background. I begin with the rough drawing, crawling from one area to the next. I begin washing in the background, which will require forty layers. Finally I bring out the amazing highlights of the lantern fish. I respect the transparency of watercolor and believe the medium is at its best when layered. But often I build on a rich surface with what I refer to as "late opaques." I am not using the special watercolor paint called gouache, known for its ability to cover, but rather more typical translucent watercolor with very little water. Lemon yellow (nickel titanate) is an example of a watercolor that will appear opaque and actually lighten areas that have gotten too dark. All the red, yellow, and orange cadmium watercolor paints are also opaque. The less water you mix with the paint the more opaque it is. By opaque I mean it is less transparent and blocks the colors that have gone before. The use of transparent washes is a strength of watercolor. With layering, many colors can show through and combine in a unique way to make a final color. The strength of the cadmiums, when used nearing the end of a painting, is that they have a unique glow and power over all that has gone before.

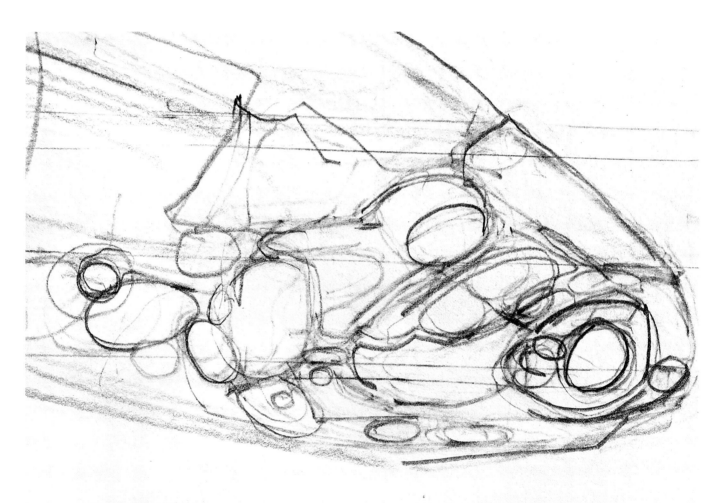

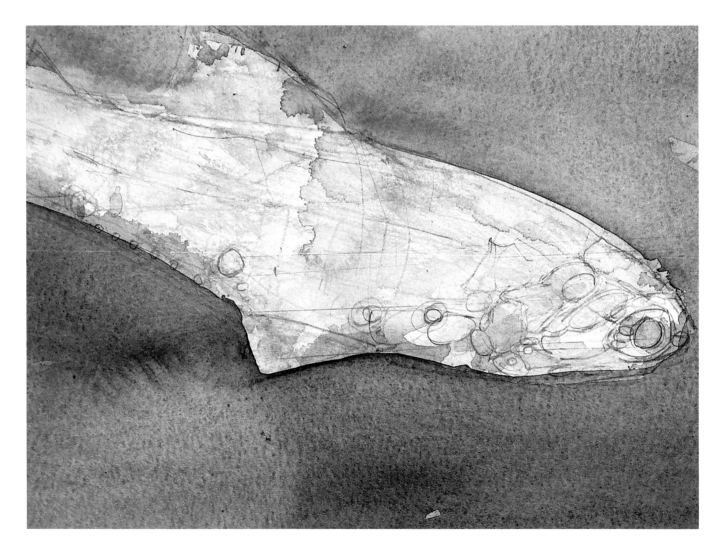

CHAPTER 14

Hydrothermal Vents

Stretching roughly from pole to pole, mid-ocean ridges cross deep-sea ocean basins. The ridges aren't necessarily in the middle of the ocean basins, but they are ridges, elevated to around 1.5 miles (2,500 meters) above surrounding plains that are at depths of nearly 2 miles (3,000 meters). I've been on fourteen cruises to explore hydrothermal vents using *Alvin* or Remotely Operated Vehicles with geophysicists, geochemists, and other biologists. I've seen chimneys 130 feet (40 meters) tall and clefts in the seafloor spewing cloudy fluids colonized by unique fauna and densities of animals that exceed anything I've ever seen on land—or imagined. These are special.

Hydrothermal vents sustain high densities of animals because their energy comes not from plants using sunlight (there is no sunlight at these depths) but from bacteria. Bacteria use chemical reactions and material in the cloudy vent fluids to get energy. In

Grey smoker

technical terms, plants use photosynthesis and bacteria use chemosynthesis. At mid-ocean ridges, heat rises from deep within the earth to near the surface of the seafloor. The heat warms the surrounding basalt that forms the seafloor, causing it to expand and elevate the ridge. The basalt cracks, allowing seawater to enter under the seafloor. As this water is heated, it rises, drawing more water under the crust. The heated seawater, as it moves through the basalt, carries away some of the chemicals (mostly metals) that form the basalt. In addition, the heat and pressure under the basalt chemically alter some chemicals in the seawater. One of those chemicals, sulfate, is converted to sulfide—a toxin. The sulfide couples with the dissolved metals to contribute to the hydrothermal vent fluid. Leaving the basalt, this fluid may exceed 750 degrees Fahrenheit (400 degrees Celsius); it doesn't boil because of the high hydrostatic pressure. The fluid cools quickly once free of the seafloor—the surrounding water is about 36 degrees Fahrenheit (2 degrees Celsius). The toxic, heavy metals the fluid contains are

Grey smoker and giant tube worms (*Riftia pachyptila*). Length to over 9.8 feet (3 meters).

foul to us but are food to the bacteria. Some bacteria form thick mats on rocks that are exposed to fluid flow; some live mostly suspended in the fluid; and some live inside animals, forming symbioses—partnerships between organisms living in close contact and helping one another, like the coral and algae we met earlier.

The most famous vent symbiosis among animals may be tube worms and their bacteria. Once described as a distinct phylum (as different from other animals as you can be and still be an animal), tube worms are now considered to be polychaetes and members of a family named more than a hundred years ago for tiny worms with tubes, the Siboglinidae. The larvae of tube worms look like they'll grow into perfectly normal segmented bristle worms, or polychaetes, but when they eat the right bacteria, they change. They settle, and as they metamorphose into their adult form, their mouths and anuses close up, forming a bag around that bacterium that begins to divide to make many copies of itself. For the rest of their lives, the worm and bacteria are together, mutually dependent. The worm builds a tube around its body. The tube supports the large plume on its top. Through the plume, the worm absorbs dissolved oxygen and sulfide that it then delivers via its bright red blood to its bacteria. The bacteria convert those dissolved gases to other forms and harvest energy in doing so. They share the energy with the worm and they both thrive.

Other animals also have symbiotic relationships with vent bacteria. Limpets—snails that have a more or less conical shell rather than a coiled shell—are sometimes coated in bacterial mats; some species host bacteria in their insides. These limpets don't have to eat (the bacteria feed them as happens in tube worms), so when these limpets find a great place for their bacteria, they live stacked on top of each other. Clams, including the beautifully named *Calyptogena magnifica*, extend their foot (clams only have one) deep into cracks between rocks to reach wafting vent fluids that are out of reach of other vent animals. These animals are specialized to provide a good home for their life-sustaining bacteria. This means that the

animals' survival depends on vent fluids. Vent fluids, however, are prone to clogging their conduits because they carry a lot of metals, because the rocks can collapse in microearthquakes, or just because they grow to be too heavy. The fluid finds another way around the clog, but the animals at the original outflow may be stranded—only metaphorically high and dry—in cold water without life-sustaining nutrients. If they can't move far enough to find the new orifice, they die. Only one vent animal, the mussel, can filter feed. This allows these animals to survive, sometimes for long enough for the fluid flow to return.

As the hot, metal-rich vent fluid leaves the seafloor and the vent, it suddenly meets cold seawater and is instantly cooled. The metals in the now-cool vent fluids are supersaturated; they precipitate and fall to the seafloor. It's like making rock candy: the sugar you've added to the hot water crystalizes when the mixture cools. Where fluid flow is sustained, the metals pile up. Chimneys made up of these metals can be 130 feet (40 meters) tall. Animals, of course, can't live where the superhot fluid is escaping, but they get very close. Animals like *Alvinella pompejana*, the Pompeii worm, live on the outside of chimneys. These worms exist in the critical boundary between the superheated fluid and the surrounding seawater.

It's a dicey arrangement. Chimneys can fall, and fluids can change course, starving or boiling animals. What makes it even dicier is human's increasing need for rare metals—found in relative abundance in chimneys—for technology, our phones, computers, and batteries for electric cars. Some companies are betting that there is money to be made mining deep-sea vents. Inconceivably huge machines are in the "preprototype" stage, designed to grab vent material, maybe whole chimneys, grind them up, then convey the residue to the surface ship and return the "debris" to the deep sea. How far the debris, with its toxic metals and sulfides, will spread in the deep sea is unknown. How deep-sea animals, which live in the most constant uniform habitat on earth, will react we don't know. The impact may be disastrous.

ARTIST NOTE

The hydrothermal vents were quite a challenge. I would build the impression of smoke by applying the paint and then lifting in the lighter areas. This didn't work so I used two paints—titanium white mixed with manganese blue—to imply a softer lit area. I kept going back and forth. Here is where that heavy (300 pound) paper is necessary. I am always moving into unknown territories. And rather than look to see how others do it (I have never had a watercolor class), I just try things out. So my suggestion to the reader is "just do it."

Keeping soft images was paramount. You can soften a hard edge with water or those paints that I mentioned—titanium white and manganese blue. The puffs of smoke needed to have a lit side as well as a shadow side. This required persistence. I think this is the crux of how I use watercolor . . . I create form gradually with no clear sense of where I am going until I get there!

CHAPTER 15

Food Falls

Food is scarce in most places and for most of the time on the deep-sea floor. It becomes hyperabundant locally when a large food packet falls from the overlying waters. The most famous examples are dead whales. When whales die, their bodies float on the surface as they putrefy; their blubber, gases generated by rot, and any air in their lungs give dead whales buoyancy. Bacteria, sharks, and big fishes feast on the remains, removing any soft tissues they can reach and eating all they can. Eventually the weight of what's left makes the body sink. When the body of one of these leviathans hits the bottom, other scavengers like small amphipods and good-sized hagfish feed on the increasingly soft tissues until all that remains are the bones.

Despite all the whales that have died over time, the seafloor isn't littered with whale skeletons. Bone-eating worms (*Osedax* spp.) bore into the bones and dissolve them. Despite their name, these worms don't

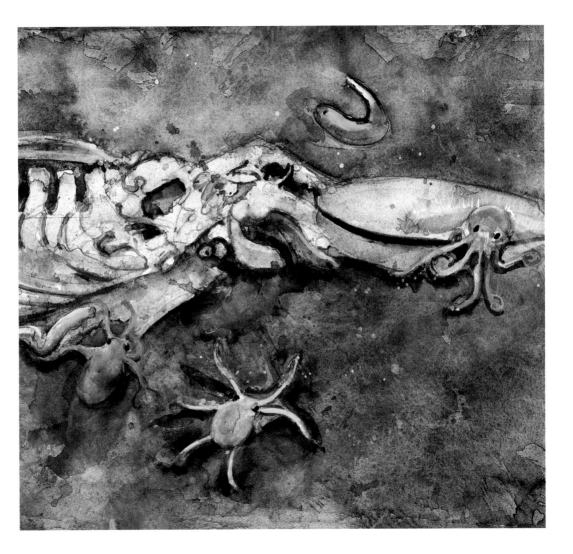

actually eat the bones directly. They live on chemical energy that is harvested by bacteria living symbiotically in the worms. Their bright red plumes extend away from the bones to grab oxygen needed by the animals, and the chemical reaction of mixing that oxygen with sulfide from the rotting bones sustains their bacteria, which in turn, sustain them. This is very similar to the way tube worms make a living at hydrothermal vents.

Dead whales aren't the only large food parcels to fall from above. Trees sometimes fall to the ocean floor. Of course, trees don't grow in the ocean, but they can wash into the sea from banks along a stream or river and drift in the currents until they sink. Why do they sink? Animals like gooseneck barnacles colonize them as microscopic larvae; the barnacles filter feed and grow large and heavy, as do other animals that attach to the trees. Animals that are late settlers are fewer in number, as their larvae must pass through the veritable death trap created by the filter-feeding barnacles, which reach into the water column to grab such choice morsels as a larva ready to settle. The tree also absorbs water as it floats. The water pushes the air in its vessels out as time goes on and makes the tree heavy enough to sink.

When the tree hits the seafloor, it must stir up clouds of sediment. Among the first animals to colonize the sunken tree may be the most important, wood-boring bivalves. These animals live only on, or rather in, wood; with the help of bacteria they can digest the wood and convert its energy and nutrients into forms other animals can use. Even in the middle of the ocean, larvae of these clams can be common. They settle on the wood and immediately undergo metamorphosis: from swimming larvae, they turn into bottom-dwelling bivalves. A wood-boring bivalve, however, is different from other bivalves. A wood-borer's shells have teeth and don't close completely—they gape to allow the foot to stick out. The foot of each one attaches to the wood almost like a suction cup; it stabilizes the bivalve as its two valves separate then come together, scraping against the wood with their teeth as they do so. Bit by excruciatingly small bit, the clam bores a hole into the wood. As the hole deepens, the clam descends into it; forever after only the tip of its siphon will be exposed. The bivalve spends the rest of its life pulling its valves apart, then closing them to scrape the wood so it can eat the shavings.

How do the larvae of these animals find sunken trees or even sunken twigs? How much wood is out there? These are good questions. Until we know more about what happens on the seafloor in general, we won't know more about these rare events that sustain not only these wood-boring clams and bone-eating worms but the host of animals that are known only from food falls—immense banquets of energy and nutrients—in the deep sea.

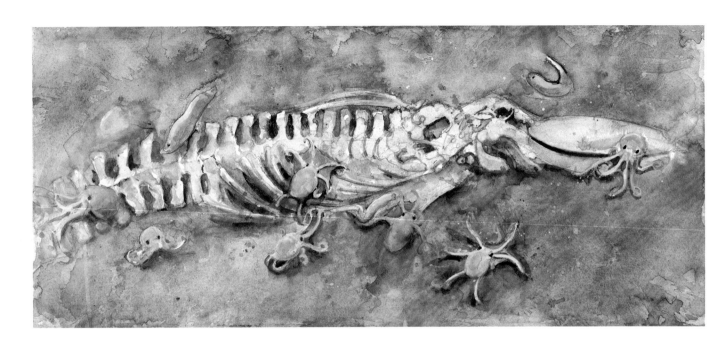

Octopus (*Muusoctopus robustus*). Total length nearly 14 inches (35.5 centimeters) long.
Zoarchid fishes or eelpouts Family Zoarcidae (over 300 species). 5–43 inches (13–110
centimeters) long.

Building a rich background is the result of many, sometimes fifty, layers of paint. The trick is to let each layer dry completely. If it is still cool to the touch then it is not completely dry and will lift up when the next layer is put on, defeating the purpose. When you think "I must be done by now" you probably are not done. Don't worry about covering up the images with a layered background. Cover the images with paint in order to prevent edges from building up. Then lift the paint from the image areas with a paper towel.

CHAPTER 16

Deep-Sea Floor

As the submersible sinks below the bioluminescence of the deep scattering layer, it enters an apparently mostly empty area of the world's oceans. Small bits of food falling from above have been recycled through the animals living in the upper layers of the water column. The food has now been reduced to what is called "marine snow," which is descriptive. It looks like flakes drifting down without end. Each time an animal eats a flake, it takes whatever energy it can from the flake and poops out what remains. Those remains drift down to the animals below. Finally, the flakes settle on the seafloor, where they accumulate over eons.

The fine sediment on the bottom is shifted by currents that in places can be strong enough to leave ripple marks on the bottom. Where currents are weak, the material builds up as fluffy sediment. If you could move your hand over it to generate a wave, the sediment would waft up in the water column and then oh so slowly settle back to the bottom.

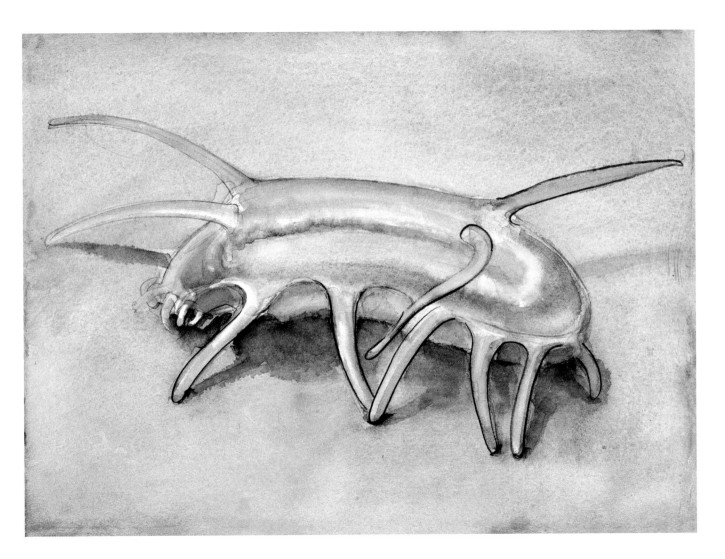

In the midst of the fluffy sediment live small animals that basically no one has ever seen. They have to be tiny, as there isn't enough food for them to be large. But they can and do live in big patches with many different species. That's the odd thing. These individuals live among animals of different species; rarely do they live near an animal of their own species. How do they find a mate? How do they get to a mate once they find one?

Even though there isn't much food, hungry predators come by. Big ones. They don't have huge gnarly teeth because when they eat the tiny animals in the sediment, they don't have to chew. The predators do, however, need to travel from one patch with a lot of these tiny animals to another such patch while spending very little energy. How? They swim, of course. Dumbo octopuses use their gigantic fins and web to move across the abyssal plain. Warty octopuses must walk between patches as they don't seem to swim very well. I think these octopuses continually run their arms through the sediment as they walk and use their suckers to grasp little animals, then move them to their mouths.

Sea cucumbers in shallow water are descriptively named. Unlike the vegetable, they crawl on small tube feet as they eat sediment, digesting the good parts of it and pooping out the rest. On the seafloor these animals have mostly eliminated their tube feet and elaborated their bodies, often transforming the parts over their heads into a set of what look like wings. The bodies of deep-sea cucumbers aren't solid like a garden-grown cucumber; they are sheer. They swim by flexing their bodies from head to tail to generate lift. As they rise off the seafloor, their sediment-filled gut shows through.

Rat-tail fishes also patrol these depths, ever ready to scavenge on the bodies of large animals that died

Opposite, page 117: Warty octopus (*Graneledone pacifica*). Up to about 40 inches (1 meter) long.

Page 118: Swimming sea cucumber (*Enypniastes eximia*). 4.3 to 9.8 inches (11 to 25 centimeters) long,

Page 119: Warty octopus (*Graneledone* sp.). Up to about 40 inches (1 meter) long.

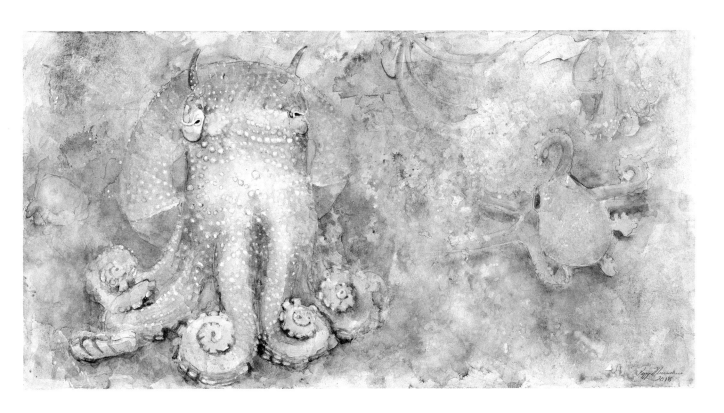

some time ago but only now are sinking to their final repose. Rat-tails roam across huge distances, led on by tantalizing odors of decay that perhaps only they can smell. Tiny remnants of those large rotten bodies, scavenged of meat and decaying to near nothing, might fuel those rich patches of tiny, tiny animals impossibly far away from the coast and miles under the sea surface that we are just learning about.

ARTIST NOTE

I'm not sure of much when I begin a piece. But I don't sit back and "think." Rather, I put something down. I respond to visuals, not thoughts. So I began this piece with a single image of the sea cucumber. Next I tried to make it move. But I am still an amateur animator. My students do it better. Its transformation was too abstract, so I washed away my attempts. But this didn't tell the story of its elegant movement. So I started adding related species to show variety. Being wrong isn't the end. It leads you to the right.

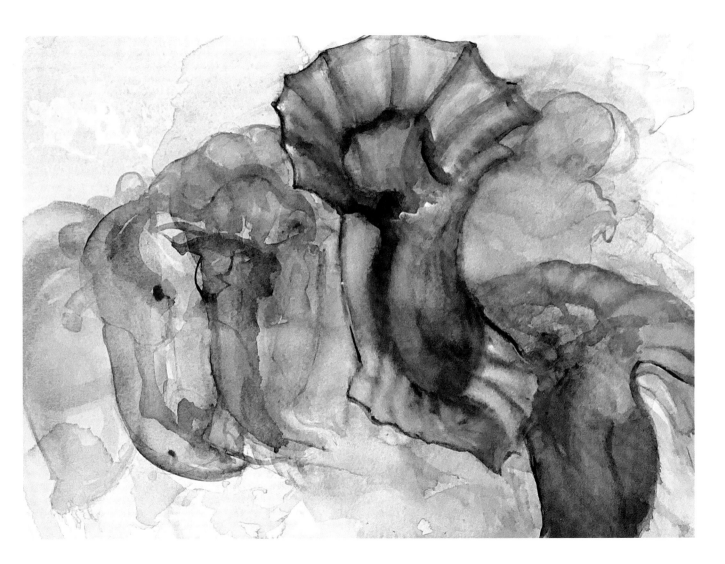

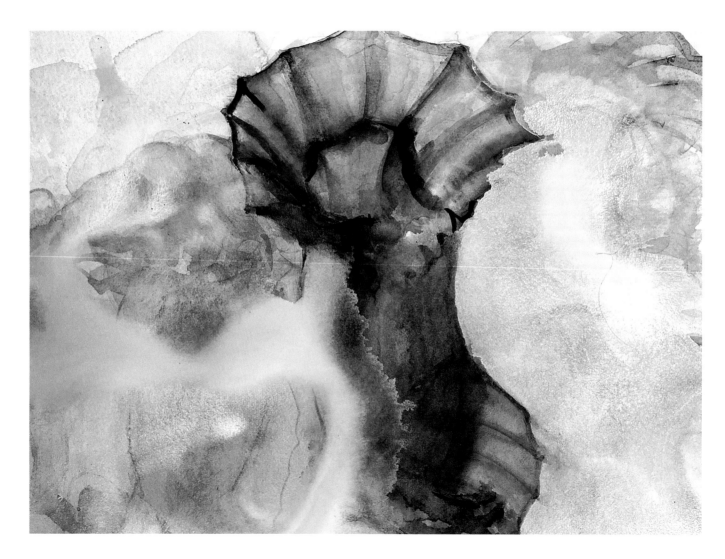

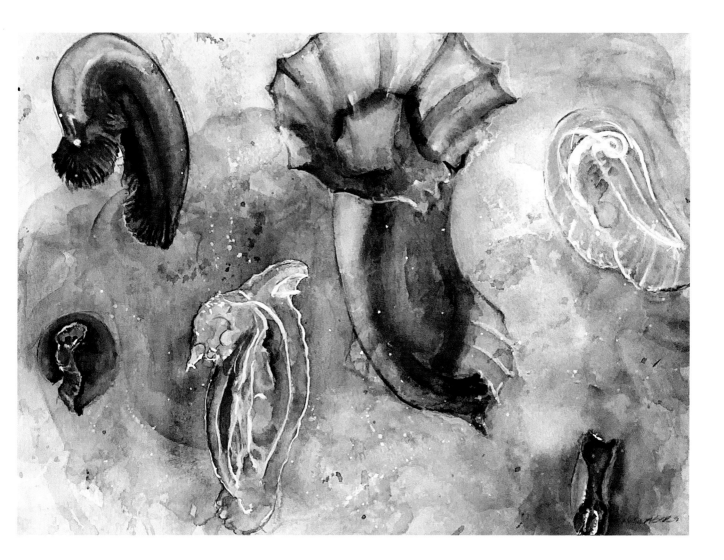

Epilogue

All the habitats of the oceans are connected to each other. They are also connected to the land. You might still think that the deep sea is the final resting place for all the energy and nutrients produced on land, that once they enter the depths, they are lost forever, just as we hope the garbage that has been dumped there through the preceding millennia will never be seen again. However, the deep sea, or parts of it, do reemerge here and there.

Across the millions of years of geological time, forces deep inside the earth push deep-sea sediments toward the continents. Hot magma under the seafloor at mid-ocean ridges causes the ridges to bulge, rising to shallower depths than the surrounding seafloor. This magma expands the rock and it rises—and sometimes it erupts, covering the seafloor with hot lava to form the newest real estate on Earth. These new eruptions push the remnants of older eruptions, which are now cooler and denser, aside and away from the hot ridges. This basalt, formed at mid-ocean ridges, moves toward the continent about as fast as your fingernails grow, but in time it reaches them. The seafloor, being made of dense, heavy basalt, is usually pushed under the comparatively light, thicker continents that you could almost say float, but in some places bits of the seafloor are pushed up, to be found on land.

This whole scenario may sound pretty far fetched, but these former seafloor deposits have had a big impact on humans. Distinct geological structures called ophiolites are probably the most important, as they are often associated with metal-rich deposits from ancient deep-sea hydrothermal vents. The copper ore from ophiolites on Cyprus has been mined since the Bronze Age. As copper is a key mineral in bronze, those ophiolites may have contributed to the advance of civilization.

Our interconnections with the deep sea continue today. Some deep-sea squid females rise to shallow waters to release their eggs and young. Shallow water, which is warmer and offers more food, is an ideal nursery for baby squids. The warm productive waters speed the growth of these tiny squids, making them bigger faster. Bigger is safer, but some are still eaten, giving the deep-sea nutrients their mother packed into their eggs to the shallows. As the survivors grow, they begin to move deeper, taking with them the nutrients they gained from above. Those squids may move through all the ocean habitats in the course of their lives. The transit of nutrients in the oceans, then, isn't just from the top down.

As humans, we are just starting to learn more about the oceans and their depths. The oceans cover 70 percent of our planet. They shape our climate and lessen the effects of climate change. They offer food, oil and gas, minerals, and untold biodiversity. Even without those resources, we're connected with the depths of the ocean on more than a spiritual level. We're part of them, and they need our protection.

FINAL ARTIST NOTE

There is more to be explored, and I began this underwater adventure with leafy sea dragons. These I chose just for their uniqueness and relation to habitat. So I'll leave you with these images and encourage you to make your art an adventure.

Leafy seadragons (*Phycodurus eques*). About 14 inches (35 centimeters) long. *Opposite:* Seahorses (*Hippocampus* spp.). 46 species. 0.6 to 14 inches (1.5 to 35 centimeters) long.

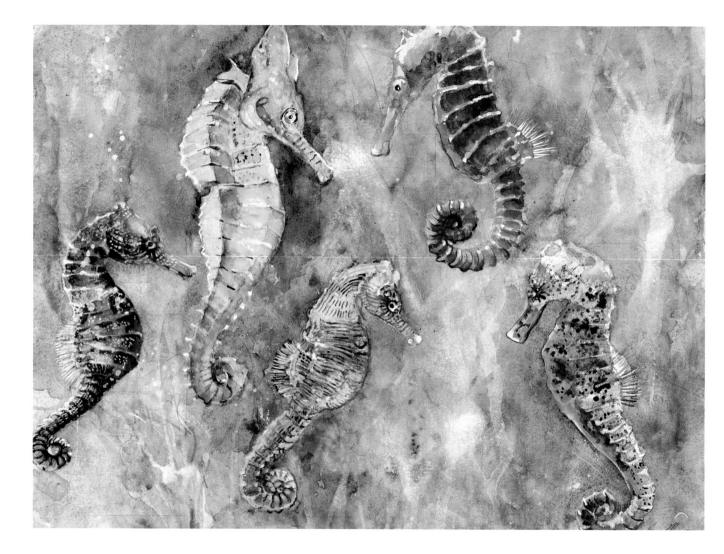

Acknowledgments

Janet Voight. Thanks to Michelle Flitman for her extensive work on the book's early design. Meriel Brooks and Kevin Feldheim were very helpful in identifying fishes and the shark in the paintings. To people with whom I have been to sea—the subsea vehicle pilots, the ship's crews and captains, and all the scientists who taught me something new—you all deserve much more praise than I can give here. Thank you so much.

Peggy Macnamara. Many thanks to Paul Lane, who photographed all my paintings and made many adjustments when needed. And to my daughter Katie, who edited in her own fashion with expertise. To Janet for her patience and humor. To the zoology department at the Field Museum, which has given me a home for the past thirty years.